SpringerBriefs in Materials

More information about this series at http://www.springer.com/series/10111

Prashant Jindal

High Strain Rate Behavior of Nanocomposites and Nanocoatings

 Springer

Prashant Jindal
University Institute of Engineering
    and Technology
Panjab University
Chandigarh
India

ISSN 2192-1091   ISSN 2192-1105   (electronic)
SpringerBriefs in Materials
ISBN 978-3-319-14480-1   ISBN 978-3-319-14481-8   (eBook)
DOI 10.1007/978-3-319-14481-8

Library of Congress Control Number: 2014959262

Springer Cham Heidelberg New York Dordrecht London

Springer International Publishing AG Switzerland is part of Springer Science+Business Media (www.springer.com)

Contents

Chapter 1
Introduction

The success of any equipment used in an engineering application is dependent upon several factors. One of the most important decisions to be made for designing any engineering equipment or a device is the selection of fabrication material of the device. The choice of the fabrication material is dependent upon the application and the operating conditions where the device has to be deployed. These conditions can be dependent upon pressure, temperature, load, speed, medium, current, resistance, flow, etc. Apart from these, there are many other commercial factors which need to be considered such as safety, costing, reliability, and size. Keeping all these functional and commercial parameters under consideration, one parameter which is core in influencing each of these factors is the choice of material for fabricating the device. Therefore, investigation of material behavior under such variable conditions becomes critical. Due to the vast choice of materials, material selection to suite all the parameters in itself becomes very important.

Modern day engineering applications are varied. A single device has to operate under extreme and varied conditions, and during these conditions, the same device has to perform different types of functions. To sustain under all these factors, an engineer has to make an extensive study on the type of material he is going to use for the development of such a device. Depending upon the variety of applications and operating conditions, engineers can select a single material, combination of two, or even more than two materials. Even if he selects a material that is suitable for only one kind of operating condition, he can still use that same material but in combination with some other suitable material which can complement during adverse operating conditions. Hence, the combination of these two materials determines the success of that particular device under the specified working conditions.

© The Author(s) 2015   1
P. Jindal, *High Strain Rate Behavior of Nanocomposites and Nanocoatings*,
SpringerBriefs in Materials, DOI 10.1007/978-3-319-14481-8_1

1.1 Composites and Coatings

Compounds, alloys, mixtures, etc., have been developed to fabricate different types of devices to sustain under variable working conditions and applications. For building massive structures, steel has been used successfully for many decades, which is an alloy of iron and carbon. But even iron and carbon separately are used extensively for many other applications and different operating environments. However, with the passage of time and new challenges in the scientific modern science, lots of new materials have come up. Among these, new composite materials have gained a lot of attention. Fabrication of composite materials involves combination of two or more materials in such a way that the original component remains unaffected. The base material is known as matrix, and the added filler material in smaller proportions is called reinforcement. This is, basically, done for the purpose of taking advantages of the material properties of reinforcement and matrix material, so that their combination generates more useful properties. If one of the constituents is a particle of nanometer (10^{-9} m) size, then it is called a nanocomposite. If the aspect ratio of the reinforcement is large, then it overcomes the low modulus of the matrix because aspect ratio is responsible for higher load transfer. Similarly, the overall toughness of the matrix is dependent on the ultimate tensile strength of the reinforcement. Due to the use of nanomaterials, maximum strengths are achievable [1] in the composites as the mechanical properties become insensitive to flaws at nanoscales.

The challenge in developing nanocomposites always remains on how to exploit the unique properties of the small-sized constituents to a bulk or macroscopic scale. Composites that have gained lot of popularity have used a plastic or polymer form as the matrix. In a plastic-based matrix, reinforcement can be of many types. Generally, plastics are soft, moldable, easily available in bulk available, light in weight, etc., which makes them a suitable base material in a composite. On the other hand, reinforcement material can be a material which is available in small-sized quantities and has immense strength, high conductivity, and any other special structure-dependent property. However, creating functional and structural multi-phase nanocomposites is still under development stage for ceramic and polymer matrices.

Another way to use devices is by covering/coating the base material by some other material which has certain desirable properties such as conductivity, absorption, reflectivity, strength, and cushion effect. However, such devices can be applied to very specific applications where protection of the base material from any external effect such as scratch, high or low temperature, shock pressure, impact load, and heavy fluid flow. is required. In case, the base material has to be protected from any impact or any shock transfer from the environment that needs to be reduced, then a specific type of covering can be done on the base material. Choice of base material and covering material and interaction between the base and cover become an important parameter in such situations, so that the coated combination material is able to perform the desired function without any obstacle.

1.2 Carbon Nanotubes (CNTs) as Reinforcement and Coating Material

In the recent past, carbon nanotubes (CNTs) have gained a lot of attention because of their remarkable physical and mechanical properties. Ever since the synthesis of CNTs [2] and study that followed exploring mechanical and structural properties of CNTs [3–6], there has been wide-ranging interest in the scientific and engineering communities to exploit these for varying applications [7, 8] such as pressure sensors, electrodes in batteries, conductive films, superconductors, heat radiators, nanotube membranes, and shock dampers. The calculation and measurement of various mechanical properties such as elastic constants and bulk moduli of CNTs [5, 9] have revealed their tremendous strength and resilience. The unusual mechanical strength of the CNTs revealing them as about 100 times stronger than steel motivates us to fabricate and modify useful materials which are cheaply available in bulk form by embedding CNTs in various forms to make composites which have the desired mechanical properties. In fact, it is not only the static properties such as hardness, tensile, and compressive strength but also the dynamical properties like impact strength that are of great significance. In addition to this, the possibility of using CNTs in energy and shock absorption applications as a protective layer is of great importance. Therefore, CNTs because of their unique properties seem to provide a means to achieve both these objectives either by making nanocomposites or by coating.

1.2.1 Structure of Carbon Nanotubes

The CNTs are an offshoot of fullerenes which are have round or semi-round cage structure. Fullerenes were discovered in 1985 while studying spectra of interstellar space. The most abundant of these fullerene molecules is also called carbon C_{60} because this is formed by 60 carbon atoms lying on the vertices of a truncated icosahedron.

Subsequently, CNTs were discovered in 1991 by the Japanese electron microscopist Iijima [2] while studying arc-evaporation synthesis of fullerenes.

CNTs consist of graphitic sheets seamlessly wrapped to cylinders. Graphitic sheets are single layers of graphite. With only a few nanometers in diameter (1–50 nm), yet grown up to millimeter length, the length-to-width aspect ratio of CNTs is extremely high. They can exist as single-walled carbon nanotubes (SWCNT), concentric or multi-walled carbon nanotubes (MWCNT), or randomly arranged form as a CNT bunch. Single-walled nanotubes are a very important variety of carbon nanotube because they exhibit important electric properties that are not shared by the MWCNT. SWNTs are the most likely candidate for miniaturizing electronics past the microelectromechanical scale that is currently the basis of modern electronics. MWNTs consist of multiple layers of graphite rolled on themselves to form a tube shape. The structure of fullerenes and CNTs is shown in Fig. 1.1.

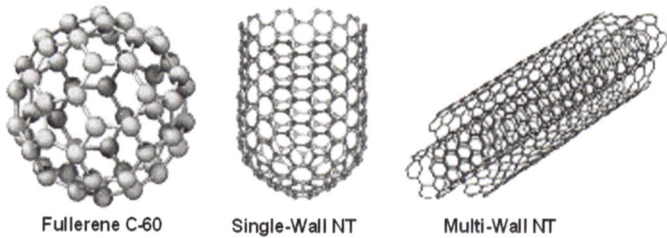

Fullerene C-60   Single-Wall NT   Multi-Wall NT

Fig. 1.1 Different types of caged structures (http://www.hzdr.de)

Fig. 1.2 A typical graphene
sheet, showing Armchair,
Zigzag, and Chiral Vectors.
The basic vectors defining a
chiral tube are shown as b_1
and b_2

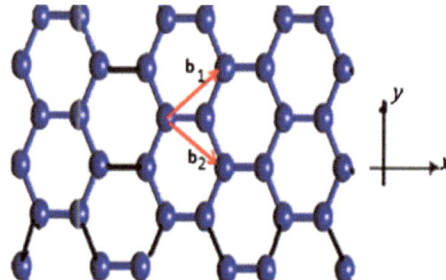

The interlayer distance in multi-walled nanotubes is close to the distance between graphene layers in graphite, approximately 3.34 Å.

There is a special place of double-walled carbon nanotubes (DWNTs) which must be emphasized here because they combine very similar morphology and properties in comparison with SWNTs and improve their resistance to chemicals significantly. This is especially important when functionalization is required (this means grafting of chemical functions at the surface of the nanotubes) to add new properties to CNTs. In the case of SWNT, covalent functionalization will break some C=C double bonds, leaving "holes" in the structure on the nanotube and thus modifying both its mechanical and electrical properties. In the case of DWNT, only the outer wall is modified. DWNT synthesis on the gram-scale was first proposed in 2003 by the chemical vapor deposition (CVD) technique, from the selective reduction of oxides solid solutions in methane and hydrogen.

A typical graphene sheet is shown in Fig. 1.2.

1.2.2 Features and Properties of Carbon Nanotubes

These nanotubes are the strongest fibers known [7, 10]. A single perfect nanotube is about 10–100 times stronger than steel per unit weight. The Young's modulus of CNTs which is nearly 1 TPa is a crucial property to be considered for mechanical engineering applications; however, the modulus of these materials can vary

Table 1.1 Various mechanical properties of CNTs

S. no.	Property	Experimental values	Theoretical values
1	Young's modulus (TPa)	0.5–2.5	0.5–1.24
2	Bending strength	15 GPa	–
3	Tensile strength	13–52 GPa	–
4	Shear strength (MPa)	0.08–0.3	0.66
5	Poisson's ratio	–	0.19–0.22
6	Thermal conductivity (at room temperature)	3,500–6,600 W/mK	–
7	Density	–	100 mg/cm^3

in tension and compression. Not only are CNTs extremely strong, but they have very interesting electrical properties. A single graphite sheet is a semimetal, which means that it has properties intermediate between semiconductors (like the silicon in computer chips, where electrons have restricted motion) and metals (like the copper used in wires, where electrons can move freely).

They also have unusually large thermal conductivity [11, 12]. The behavior of any material when subjected to a change in temperature (heat) is termed as the thermal property of the material. All nanotubes are expected to be very good thermal conductors along the tube, exhibiting a property known as "ballistic conduction," but they are good insulators laterally to the tube axis. It is predicted that CNTs will be able to transmit up to 6,000 W/mK at room temperature, and comparing this with copper, a metal well-known for its good thermal conductivity transmits only 385 W/mK. The temperature stability of CNTs is estimated to be up to 2,800 °C in vacuum and about 750 °C in air [13].

Low density and other physical properties make CNTs a very attractive prospect for mechanical engineering applications such as developing strong materials, shock/pressure absorbers, and modifying heat transfer ability of substances in the form of a filler in a nanocomposite. Some of the important mechanical properties [11, 13–20] of MWCNTs have been listed in Table 1.1. Due to such attractive properties and applications, there is a great interest in finding out more methods to evaluate and validate these properties by using various experimental and theoretical techniques.

The atomic lattice structure of nanotubes is associated with a high mechanical stiffness as well as great flexibility. Because the nanotubes are highly symmetric molecules linked by many covalent bonds which are in a parallel cylindrical configuration, nanotubes are both very flexible and strong. Regarding the flexibility, nanotubes can be bent strongly which buckle [21]. Small-diameter SWNTs can be elongated by ~30 % before breaking [22, 23] and the value for the breaking strength is 55 GPa [24]. Together with their high aspect ratio, this makes them ideal candidates for tips of scanning probe microscope like the atomic force microscope (AFM). In the macroscopic world of plumbing and drinking straws, we are used to the fact that buckling of cylindrical structure leads to induced dislocations or even fracture. On that scale, buckling is thus not reversible. By contrast, buckling of CNTs is reversible. The influence of buckling on the electrical properties has very interesting surprises [25].

With the passage of time, lots of techniques for production of CNTs have also come up and now plenty of CNTs are produced every day; however, fabrication of nanocomposites with CNTs as reinforcements remains a challenge. Methods to process such composites are still in development stage.

Dispersion of these nanomaterials in the composite remains a difficult task as without proper dispersion uniformity in property in the composite material is not achievable. Particle–particle and particle–matrix interactions play a crucial role in predicting the overall behavior of the material under any operating condition. So dispersion is very important in producing composite materials. But still, considering a goal to achieve an overall improvement in the composite property, complete dispersion and evaluation of property at the particle level become secondary. Even if mixing certain amount of nanofiller enhances the overall composite property by any amount, it should be considered a success.

1.2.3 Methods to Produce CNTs

Both MWCNTs and SWCNTs require complex methods for production in large quantities. During the process, pure nanotubes are produced along with lot of amorphous carbon and several other undesirable substances.

Broadly, the production methods can be divided into two categories: arc-discharge method which uses high-temperature evaporation and CVD [26, 27].

In the high-temperature methods, evaporation of pure carbon takes place in the presence or absence of a metallic catalyst. But CVD method always involves a catalyst. In the arc-discharge method, a DC plasma arc is produced between two carbon electrodes in an inert gas atmosphere. This arc results in production of a black fibrous deposit on the cathode which are mostly MWCNTs. Another alteration in this method can be the use of triple deionized water bath. Carbon electrodes are brought close to each other in a bath of triple deionized water, and AC of 60 A current is passed through the electrodes. This method produces black soot like material, most of which is MWCNT. Laser ablation is another similar process, used to generate the carbon vapor. Generally, it is usually used for the production of SWCNTs. A furnace at 1,200 °C consists of a graphite material which is evaporated under high-powered laser, under inert atmosphere, and in the presence of a metal catalyst [28].

The resulting products from the above methods are relatively impure containing other, unwanted impurities. However, these techniques are cost effective. But for experimental purposes, especially when composites need to be fabricated, large quantities of CNTs are needed. Therefore at present, only CVD-grown nanotubes are the most preferred choice of materials for composite work in both academia and industry. This method controls the growth and synthesizes large quantities of CNTs in a direction on a substrate. Substrates are mostly made of silicon, and catalysts used are mostly nano particles of iron, cobalt, or nickel. A mixture of hydrocarbon gas, acetylene, methane, and nitrogen gas is done in a reaction chamber. Hydrocarbons decompose at about 700–900 °C at atmospheric pressure on the substrate to produce CNTs. Solgel technique is also a chemical process which uses dry silicon gel to grow aligned CNTs.

1.3 Mechanical Properties of MWCNT-Based Composites and Coatings

Nanomaterials can be used as reinforcements with polymer matrices. Polymers can be either thermosetting or thermoplastics. CNTs are often mixed directly in the liquid matrix of various thermosetting polymers. Ultrasonication process helps in better dispersion of the CNTs in the matrix, and vacuum helps in keeping the composite defect free. Very low concentrations (<10 %) of CNTs are preferred because larger concentrations lead to aggregations, and hence, dispersion in the matrix becomes poor. The high aspect ratio of CNTs assists in forming strong networks with the polymer matrix. Ultrasonication and stirring result in better dispersion, but it also shortens the lengths of CNTs. Better dispersion can assist in improving properties like tensile modulus and stiffness of the composite. Functionalization [29] of CNTs is another method to improve the dispersion, as various chemical groups such as carboxyl, carbonyl, and hydroxyl are formed at the free ends of CNTs which bind themselves with the polymer matrices. Therefore, it increases the solubility and hence dispersion of the composite. However, this also causes shortening of lengths of CNTs [30]. In case of thermoplastics, dispersion [31] is better than thermosetting polymers. Various methods are adopted for fabricating and mixing the CNTs with polymers, and large volumes of solvents are always required in which the polymer matrix forms a clear soluble solution. These solvents are generally toluene, chloroform, tetrahydrofurane (THF), or dimethyl formamide (DMF).

As per Moniruzzaman et al. [32], all preparations of CNTs form mixtures of various nanotube chiralities, diameters, and lengths along with different impurities and structural defects. These parameters vary within the sample and among samples of different batches fabricated at different places. This implies that it becomes very difficult to conduct experiments that reproduce same materials and hence impossible to compare results of materials produced by different researchers. So the same thing is valid for the composites which are fabricated with other base materials that involve MWCNTs. Broadly, the fabrication methods are classified as solution blending, melt processing, and in situ polymerization. Solution blending method is the most common for fabricating composites. High-shear forces and temperature are used for dispersion of nanotubes in polymers in the melt processing method. However, in comparison with solution blending, this method produces less effective dispersion results for nanotubes in polymers. In the in situ polymerization process, the nanotubes are dispersed in a monomer followed by polymerization of the monomers. This method enables effective covalent bonding between functionalized nanotubes and polymers.

The most important factor in the solution blending method is to identify solvent in which nanofiller and base material can form uniform suspensions. Once each constituent forms a clear dispersed suspension, both are mixed together under certain temperature conditions depending upon base material properties for certain period of time. Finally, these solutions are left to be dried and thin films of the composites are obtained. Suitable spectroscopy methods such as scanning electron microscopy (SEM), transmission electron microscopy (TEM), X-ray diffraction (XRD) can be used to view the topographical changes and extent of dispersion. Subsequently, they can be converted into specific shapes using any pressing machinery. Once the specific shapes are obtained, the composite specimen can be further deployed to any mechanical characterization instrument.

Shaffer and Windle [33] have used composites of MWCNTs with poly(vinyl alcohol). CNTs were grown catalytically to be used as filler. Composite films were made by mixing of aqueous poly(vinyl alcohol) solutions, $[CH_2CH(OH)]$ n (PVOH) with CNTs, which was followed by casting and evaporation. Volumes of each sample as mixed were same. Careful mixing of the components was necessary to prevent aggregation of CNTs. To form a stable mixture, nanotubes were covered with an adsorbed layer of polymer before it was able to interact with a significant number of other nanotubes. As a result of this, the adsorbed polymer stabilized the nanotube dispersion and protected it against depletion aggregation caused by the free polymer.

Sennet et al. [34] used PC matrix for CNT nanocomposite. Melt processing technique was used for fabricating the composite. For dispersion, a conical twin screw extruder was used with catalytically grown CNTs and PC resin.

Cochet et al. [35] used the in situ polymerization method to prepare MWCNTs/ polyaniline composites. Arc-evaporation was used to produce nanotubes and sonication was done. Aniline monomer was added to this suspension in the presence of HCl. Finally, a composite was obtained after filtering and drying process. This way, MWCNTs up to 50 wt% compositions were fabricated.

Stankovich et al. [36] used graphite oxide with polystyrene. Graphite oxide was produced by oxidation of graphite by Hummers method from SP-1 graphite. It was dried for a week and then suspended in anhydrous DMF, treated with phenyl isocyanate for 24 h. Finally, it was recovered by filtration through a sintered glass funnel. Phenyl isocyanate-treated graphite oxide composite materials were obtained which were then formed into stable dispersions. Polystyrene was then added to these dispersions and dissolved with stirrer. A coagulated polymer composite was obtained by adding DMF into large volumes of methanol. The coagulated composite powder was then isolated by filtration and washed with methanol. Finally, it was dried under vacuum to remove any residual solvent and moisture. Before pressing into the hydraulic press for obtaining a final shape, it was crushed into a fine powder with mortar and pestle.

Zhu et al. [29] discussed about some methods that have been proposed to overcome the problems related to accomplishing dispersion which includes the use of ultrasonication, high shear mixing, aid of surfactants, chemical modification through functionalization, wrapping the tubes with polymer chains, etc. They fabricated composites of SWCNTs and epoxy resin. In their study, functionalized nanotubes were dispersed in DMF by sonication using a high-power ultrasonic processor and then in an ultrasonic bath. Thereafter, the epoxy resin was added, and the solution was stirred. A vacuum chamber was used to evaporate DMF. A high-shear mixing homogenizer was used to ensure homogeneity by mixing the SWNT/epoxy blend. All nanotube/epoxy composites were prepared using a 1 wt% load of both pristine SWNTs and functionalized SWNTs. Five dog-bone-shaped specimens were cut and polished for tensile testing. Following the same procedure described above, a control sample from pure epoxy resin was also prepared and tested for comparison.

Polycarbonate and carbon nanofibers (CNF)-based composites have also been studied [37] for their mechanical strength, and the results indicate that a 10 % concentration of CNF is sufficient to increase the tensile modulus by about 20 %.

High-density polyethylene (HDPE)-based CNT composites show an increase in Young's modulus by 6–10 % if CNT concentration by volume is increased by 0.2 %. However, ultimate stress does not show much variation for such small concentrations, and even at 0.44 vol% CNT concentration, the ultimate stress increases by only 3 % [38].

Role of dispersion and aspect ratio of MWNTs in polypropylene (PP) has also been studied by Prashantha [39] by investigating the effect on morphology, rheology, and mechanical performance. Polypropylene-grafted-maleic anhydride (PP-g-MA) was also added in the MWCNT-PP composite in small weight proportions to improve MWCNT dispersion. PP-g-MA itself comprised of about 20 wt% MWCNTs. The mechanical characterization showed that both the tensile and bending moduli and strengths of the nanocomposites increase by the addition of nanotubes, and the addition of PP-g-MA further improves these properties. However, the tensile elongation at break point decreases with the addition of PP-g-MA. For impact strength properties, Charpy testing was done and it was observed that impact strength increases till a maximum limit and then decreases for samples which had a notch. Increased stress concentration at the notch tip affects the fracture propagation which has been suggested as the reason for this increase in impact strength. But for un-notched specimen, there was a consistent decrease in impact strength with addition of MWCNTs for both (with and without PP-g-MA) types of composites. Aggregation of MWCNTs has been the reason suggested for this decrease. But overall, 2 wt% PP-g-MA tends to improve the impact properties significantly, and beyond this composition, reduction was observed.

Characterization of composites of MWNT/polyacrylonitrile was investigated by Weisenberger et al. [40]. The energy required to yield and break increased by nearly 80 % upon addition of 1.8 vol% MWNT which meant that both resilience and toughness for the composite increased manifolds.

Cadek et al. [41] reported that the presence of 1 wt% nanotubes in PVA increased the Young's modulus and hardness by factors of 1.8 and 1.6, respectively. The composites were fabricated by solution mixing method.

Qian et al. [42] reported mechanical characterization of nanotube/polystyrene (PS) composites. They produced MWCNTs samples with lengths of 15 and 50 μm. When these MWCNTs were added to PS, the elastic stiffness was found to be increased by 36 and 42 % for short and long lengths, respectively. Tensile strength for both long- and short-length MWCNT composites increased by 25 %. Theoretical calculations for the stiffness showed that results were within 10 % range of experimental results.

Andrews et al. [43] used MWNT/polystyrene composites containing from 2.5 to 25 vol% nanotubes which resulted in a consistent increase in Young's modulus from 1.9 to 4.5 GPa. Major increase in the modulus occurred when MWCNTs content was above 10 vol%. Surprisingly, at low concentrations of MWCNTs, the tensile strength decreased even below the neat polymer.

Mamedov et al. [44] prepared SWNT/polyelectrolyte composites using layer-by-layer method which resulted in exceptional mechanical properties. The tensile strength for the composites increased to nearly 220 MPa which is much greater than the tensile strength for some of the commercially used industrial plastics.

Cai et al. [45] fabricated polyimide/carbon nanotube (PI/CNT) nanocomposites with different proportions of CNT by in situ process. The friction-reduction and anti-wear capacity of the nanocomposite enhanced because of the increased the load capacity and mechanical strength of the CNT filler.

Siochi et al. [46] used a polyimide SWCNT composite and found that the tensile modulus and yield strength increased with the addition of SWCNT. The increase was high as compared to the film samples where SWCNTs had no preferred orientations but low in comparison with oriented fibers. Incomplete dispersion was the reason suggested for this low-level increase.

Li et al. [47] used ultraviolet radiation initiated polymerization to prepare a composite of polyacrylamide–carbon nanotubes (PAM–CNT). It was characterized using FTIR spectroscopy. CNTs enhanced the load bearing and anti-wear properties of these thin-film composites.

Liao et al. [48] reported significant improvements in the mechanical properties of the epoxy/SWNT nanocomposites by a 50.8 % increase in the storage modulus.

Coleman et al. [49] have compared the effect of composite fabrication methodology on the mechanical properties. They found that the in situ polymerization and solution blending techniques show better results as compared to melt processing. If the filler material (CNTs) was functionalized, then strength improved even further, indicating better binding between functionalized filler and base material. Finally, it was reported that in comparison with SWCNTs as filler material, if MWCNTs were used, then mechanical properties got enhanced in most cases, indicating that as a filler, MWCNTs is a superior material for polymer composites.

Xu et al. [50] investigated the effect of functionalized CNTs as filler in nanocomposites of poly (L-lactide) (PLLA) on mechanical and electrical properties. Polybutyl acrylate (PBA) has been utilized to functionalize MWCNTs by in situ atom transfer radical polymerization. The composite with PLLA was prepared using solution blending technique. It was found that elastic modulus increased by 94.7 % as compared to pure PLLA with an addition of only 2.98 wt% MWCNTs.

If the composite was made out of MWCNTs without PBA, then it made the PLLA more brittle due to agglomeration of MWCNTs; hence, no improvement in mechanical properties was observed. But the presence of PBA leads to better dispersion and lesser agglomeration of MWCNTs in PLLA.

Blocks of CNTs have been used to create effective and powerful pressure sensors. Taking advantage of the material's unique electrical and mechanical properties, researchers repeatedly pressurized a 3-mm nanotube block and observed its utility as a pressure sensor. A linear relationship between the applied force and electrical resistance was observed even when the pressure was increased on the block. This sensor can be used to sense minor weight changes and hence very useful for load varying industrial applications.

Mantena et al. [51] have investigated composite structures for navy ships applications. They observed the response under blast and shock load for graphite platelet reinforced vinyl ester nanocomposites (non-brominated). About 50 % increase in energy absorption was observed for 2.5 % graphite platelet/vinyl ester composite in comparison with pure vinyl ester.

The initiation of damage caused by an impact was investigated by Yang et al. [52]. Low-velocity impact tests were done using a drop weight impact tower for composite materials of $(0°, 90°)$ glass fiber-reinforced epoxy resin. It was reported that impact force required to initiate damage varies linearly with $t^{3/2}$ where t is the target thickness. This relationship was valid when test temperatures were set between 23 and 90 °C. The impact force for initiating damage varied with projectile (indenter) diameter, and also, greater the size of the projectile more is the impact force needed for initiation of damage.

Reinforcement on polyvinyl alcohol (PVA) and PMMA with few-layer graphene (FG) was also tested using a nanoindenter by Das et al. [53]. Interestingly, low compositions of FG (0.6 %) in PVA increased the modulus by about 20 % and hardness by about 50 %. For PMMA, also there was 50 % increase in hardness by reinforcing only 0.6 % FG. It has been suggested that hardness increases with effective load transfer ability of the composite, so FG having a large surface area assists in stronger interfacial bonding with the polymer matrix, and hence, load transfer becomes more effective.

Vivekchand et al. [54] have explained the use of inorganic nanowires (NW) as reinforcement in PVA to be more effective instead of using MWCNTs. Elastic modulus increased by almost two times with 0.8 % (by volume) reinforcement of inorganic NW. However, MWCNTs have a very smooth surface due to which the strength imparted by reinforcing MWCNTs is lesser than NW. The reasons suggested behind the increased modulus are a combination of large aspect ratio and high surface-to-volume ratio of the reinforcement.

Bull [55] used oxide coatings on glass and studied the fracture toughness for these coatings. Nanoindentation and energy-based methods were used to evaluate hardness and elastic modulus for the coated materials. Young's modulus for 400 nm thick coating of SnO_2 was 133 GPa, ZnO 117GPa, and TiO_xN_y 122 GPa. However, it has been emphasized that the size of coating does not affect the hardness and Young's modulus for each of these coating materials. In the thin coating range of 100–400 nm, all these oxide coatings have almost same hardness and modulus. The suggested reason behind this is that all the deposited oxide coatings are poorly crystalline and scale-dependent plasticity mechanisms which require the presence of dislocations on well-defined slip planes were not present in the amorphous layers.

Ehsani et al. [56] have shown the effect of glass flake (GF) and epoxy vinyl ester resin on the thermal and mechanical behavior of composites. When GF is incorporated into a coating at an appropriate level, the flakes align parallel to the substrate surface, producing a shield or barrier of overlapping plates. Specially prepared hammer milled GF made out of extra corrosion resistant glass having high aspect ratio makes the base material more corrosion resistant. Different compositions of GF were obtained using mixing method which was very effective method as it had a strong effect on the surface morphology, size, and distribution of glass flake. In case of epoxy vinyl ester resins, certain fillers such as carbon black, carbon fibers, and glass when added enhance properties like modulus and thermal conductivity.

Gao et al. [57] have presented on how surface defects cause the measured tensile strength of glass and other brittle materials to be significantly lower than their theoretical values. These surface defects can be healed by using coatings which can heal the flaws and modify surface properties. A hybrid coating layer based on styrene–butadiene copolymer with multi-walled carbon nanotubes (MWCNTs) with or without nanoclays act as a mechanical enhancement and environmental barrier layer when applied to alkali-resistant glass (ARG) and E-glass fibers. The work indicated that nanostructured and functionalized traditional glass fibers show remarkable improvements in both mechanical properties and resistance to environmental corrosion. Low fractions of nanotubes (0.2 wt% in sizing) increased the strength of healed glass fiber by 70 %.

Grujicic et al. [58] used polyurea coatings for impact absorbing applications. Application of polyurea coatings can substantially improve blast and ballistic impact resistance due to which buildings, vehicles, and laboratory test plates can be used successfully in defense applications. The reasons suggested behind the high impact resistance of these coatings are the transition of polyurea between its rubbery-state and its glassy-state under high-deformation-rate loading. The mechanical response of polyurea under impact conditions is dependent upon the difference between the test temperature and the glass transition temperature. When this difference is large, polyurea displays high-ductility behavior like an elastomer in its rubbery-state. But when the test temperature is closer to the glass transition temperature, the polyurea tends to transform into its glassy-state during deformation and which causes viscous-type energy dissipation. It has also been suggested that additional energy-absorbing mechanisms contribute to the superior blast protection capability of polyurea.

Nalbant et al. [59] have produced thin films of copper and zinc oxide on glass using spin coating method. During the process, they started with pure ZnO thin films and ended up with CuO by doping Cu in various percentages, ranging from 0 to 100 %. Crystal phases were achieved in all doping concentrations.

Wu et al. [60] studied the effect of anisotropically conductive film (ACF) joint under the dynamic loading of flip chip on glass (COG) and flip chip on flexible (COF) substrate. Impact tests were performed to investigate the key factors that affect the adhesion strength. Good absorption and higher degree of curing at higher bonding temperature accounted for the increase of the adhesion strength, while overheating caused overcuring of ACF and degradation at ACF/substrate interface, and due to which, there was a decrease in the adhesion strength. Rise in bonding temperature caused impact strength to increase, but after reaching a certain temperature, the impact strength started decreasing. Bonding pressure played an important role for a reliable electrical interconnect, but it did not have any significant influence on the impact strength. The fracture propagation during impact loading was found in the ACF/substrate interface (for COG packages) and in the ACF matrix (for COF packages). COG exhibited poor impact adhesion due to weak interaction of the ACF with the glass. Bonding pressure of nearly 60 MPa was required for both COG and COF. It was also found that for both COG and COF, particles and air bubbles were responsible for the weak intrinsic strength of the adhesive as they were responsible for the change in structure of ACF and also behaved harmfully in the ACF matrix during curing processes and loading. Adhesion strength of COF was found to be higher than that of COG under same bonding conditions.

Composites and coated specimen which are fabricated specifically to be used in any load- or stress-related engineering application must undergo static and dynamic load testing. Specific applications relevant to defense and shock involve rapid load increase in a short span of time, which is referred to high-strain-rate loading also. Composite materials can only be proposed for such specific engineering applications, provided they are able to resist the extent of impact or dynamic load.

Apart from being used as reinforcements, MWCNTs can be used as coating materials. For specific applications, equipments when exposed to loading conditions need to be protected from permanent deformation. Either the base equipment has to be made out of a strong material or composite or the other option is to coat it with some other material which acts as a protector or shield from the external impact. The coating may get destroyed in these cases which can later be replaced, but the base equipment remains unaffected during this process.

Glass is one such material which is widely used as window panes for vehicles. When these vehicles are exposed to certain loading conditions, glass being a brittle material becomes a weak link and gets destroyed. So if the glass can be coated with a stronger material, then the coat acts as a protector of this base glass piece, and when exposed to the impact, it gets sacrificed to absorb the external load.

Over the last decade, many substrates have been coated with a wide variety of materials using different techniques. Depending upon the application, coating material and method to coat are selected. Aguilar et al. [61] used zinc sulfide (ZnS) and copper Sulfide (CuS) thin films sandwiched in laminated glass. In this work, they report that coating of ZnS thin film of 40–80 nm thickness, applied by chemical bath deposition method on 2 mm-thick sheet of glass, improves the adhesion strength of laminated glass by about 20 %. Similarly, if a CuS thin film of 100 nm thickness is applied, then adhesion strength increases by nearly 10 % as compared to clear glass sheet. But if a CuS thin film of 100–150 nm thickness is applied over the ZnS coating, then the adhesion strength increases by about 20 % as compared to clear sheet, indicating that when both films are used together, the improvement in strength is higher. Higher thickness of CuS film is required to reduce the solar transmittance of the glazings for buildings in hot climates. But when the thickness of CuS is further raised to 150 nm, then its adhesion strength gets reduced even below the clear glass sheet by about 30 %. So mainly the combination of CuS–ZnS films can be used to enhance strength to about 12–14 MPa which is larger than the industrial standard strength (10 MPa) for glass.

Apart from coatings using different types of oxides and sulfides, MWCNTs can also be used to absorb load due to their attractive mechanical properties. MWCNT coatings can be bound with glass surface by covalent and non-covalent binding methods. In covalent binding, functional groups [62] are attached to MWCNTs by treating them with 3-aminopropyltriethoxysilane (APTES). APTES is an organosilane with the ability to create a self-assembled monolayer [63, 64]. Glass surface is treated with piranha solution (solution of 1:4 30 % H_2O_2 and concentrated H_2SO_4) which creates—OH moiety on glass surface. Afterward, the piranha-treated glass slides are treated with 10 mM solution of APTES in toluene for 3 h. This makes glass surface silanized with free-NH_2 group available for further reactions. N-hydroxysuccinimide (NHS)/ethylene dichloride (EDC) is used as a linker to covalently bind CNTs to glass surface. However, in non-covalent binding, the process is much simpler where MWCNTs are simply dispersed in a solvent and poured over the glass surface.

Once these specimens are fabricated using any suitable technique, then the process of their mechanical characterization needs to be initiated. In order to obtain the mechanical strength of these specimen under dynamic load conditions, various types of instruments like dynamic mechanical analyzer (DMA) and split Hopkinson pressure bar (SHPB) are extensively used in laboratories.

1.4 High-Strain-Rate Loading

During shock loads and bomb blasts, high-strain loads are encountered. Extremely high load is encountered by any equipment under these situations in a very short duration of time. This load can be single or multi-dimensional. Bomb blasts, bullets, shock loading, cyclic loading, etc., fall under this category. Dynamic loading is measured in terms of strain rate (per second). The stress–strain behavior of the specimen is studied to evaluate the strength of the specimen under such conditions. The material tested under such type of loading can also act as a protective or shielding unit to protect some other base equipment from destruction which otherwise gets destroyed when subjected to actual dynamic loading. Hence, to act as a shielding unit, the material must be small in size and light in weight so that its configuration does not affect the normal functioning of the base equipment. The

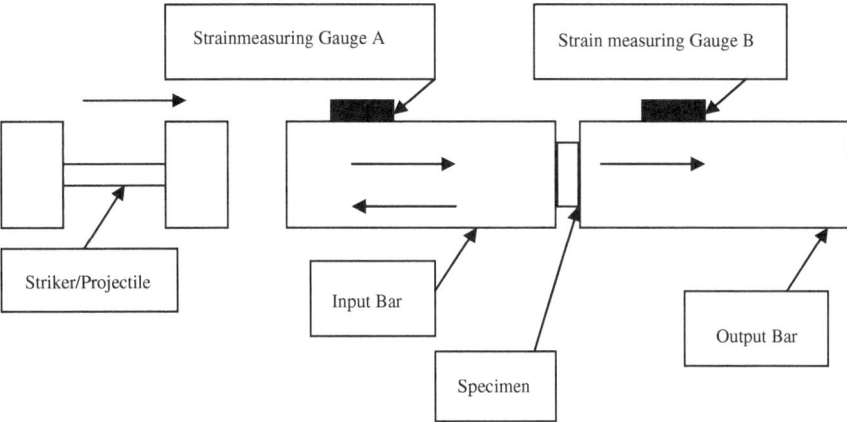

Fig. 1.3 Schematic block diagram of split Hopkinson pressure bar

materials tested under such loading can act as a single-time use unit because after one-time exposure to such loads, they may get damaged permanently. Bullet proof jackets, helmets, etc., are some of the equipments which have to be tested under dynamic loading. Split Hopkinson pressure bar (SHPB) is one such instrument which is used for dynamic load testing especially in defense labs.

SHPB is a very useful equipment to study the behavior of materials under tensile and compressive impact loading in the laboratory [65]. Stress–strain behavior of the specimen when subjected to impact or dynamic loading is obtained. Specimen undergoes a strain rate of 100–10,000s by using this instrument, and the specimen size required is in the range of a few centimeters affording laboratory testing of small-sized samples.

The SHPB apparatus consists of two long slender bars, namely input bar and output bar that sandwich a short specimen between them. A block diagram of a typical SHPB used for compressive impact loading is shown in Fig. 1.3.

High gas pressure usually acts as a source of impact which propels a projectile or a striker which is used to strike one end of the input bar. A compressive stress wave is generated that immediately begins to traverse toward the specimen. When this wave hits the specimen, it partially gets transmitted through and reaches the output bar while some part is reflected back through the input bar. Usually, an irreversible plastic deformation is caused in the specimen due to this process which lasts for less than 1 s.

The reflected pulse is reflected as a wave in tension, whereas the transmitted pulse remains in compression. The wave signal measurements are done with the help of strain gauges A (measuring incident and reflected components) and B (measuring transmitted component) attached on the input and output bars, respectively. The waves are a measure of strains which are calibrated to find stress and strain in the specimen. The details of working of split Hopkinson bar setup are widely available in the literature and the equations below form the basis of stress–strain relationships.

The incident strain (ε_I) and reflected strain (ε_R) add algebraically to transmitted strain (ε_T) as:

$$\varepsilon_I + \varepsilon_R = \varepsilon_T \tag{1.1}$$

The force on specimen (F_s) due to the impact of striker on input bar is the mean of force on the input bar (F_I) and force on output bar (F_O) as

$$F_S = \frac{F_I + F_O}{2} \tag{1.2}$$

Stress on the specimen (σ_s) is related to force on the specimen through the cross-sectional area of the specimen ($\pi d^2/4$) facing the input bar as

$$\sigma_S = \frac{F_S}{\left(\pi d^2/4\right)} \tag{1.3}$$

Force expression on input and output bars can also be written in the form of elastic modulus (E), strains, and diameter (D) of the bars.

$$F_I = \frac{E(\varepsilon_I + \varepsilon_R)\pi D^2}{4} \tag{1.4}$$

$$F_O = \frac{E(\varepsilon_T)\pi D^2}{4} \tag{1.5}$$

These equations result in relationship of stress in the specimen to the transmitted strain as

$$\sigma_S = \frac{ED^2\varepsilon_T}{d^2} \tag{1.6}$$

Similarly, the strain rate in the specimen ($\frac{d\varepsilon_S}{dt}$) is related to the wave velocity (C_0) inside the bar, reflected strain, and length (L) of the bar [65] as

$$\frac{d\varepsilon_S}{dt} = -2\frac{C_o\varepsilon_R}{L} \tag{1.7}$$

To analyze the results based on the above equations, the dynamic incident and transmitted forces must be in equilibrium.

Previous studies related to dynamic loading with large variation of strain rate on different composites such as carbon/epoxy laminate composites, Epon, and IM7/8551-7 graphite/epoxy composites have been discussed and published widely over the past decade [51, 66, 67]. Generally, dynamic strength of such materials increases with increase in strain rate. However, Hosur et al. [68] reported some deviations in this general behavior where the dynamic strength was observed to fall after certain strain rate in some materials. Additionally, effect of direction of loading, geometry of specimen fibers, angular orientation of laminates, and type of fracture for carbon/epoxy laminate composites on the stress–strain behavior have also been studied. At smaller angles of orientation of laminates, the impact strength is much higher at strain rates of nearly $1{,}000 \text{ s}^{-1}$. Laminates loaded along $0°$ possess higher impact strengths than the ones loaded along $90°$ under dynamic strain rates of nearly 800 s^{-1} [68, 69]. Chen et al. [70] worked on Epon and PMMA to find true stress–strain variation under tensile and compressive loading and showed that dynamic strength for PMMA is nearly 110 MPa for strain rate of $3{,}300 \text{ s}^{-1}$, and for Epon, it is 175 MPa at a strain rate of $2{,}500 \text{ s}^{-1}$. The results reported on tensile and compression tests on PMMA indicated that although the maximum stress increases with strain rate, the maximum strain did not consistently show increase with increased strain rate. For tensile tests, the strain rate was kept very low in the order of $10^{-4}\text{–}560 \text{ s}^{-1}$ while the compression tests were carried out in the range of $1{,}000\text{–}5{,}000 \text{ s}^{-1}$.

However, it has been stressed that more varied database is needed to have a consensus on the pattern of the results. Zappalorto et al. [71] have modeled features related to nanomodification of polymeric resins for structural applications with a perspective of obtaining high toughness even for low volumes of nanofillers. These results are related to the energy dissipation through the damage mechanisms occurring at the nanoscale. They claim that nanoparticle debonding plays an important role either as the damage mechanism itself or it acts as a trigger for phenomena like plastic void growth or matrix shear yielding.

Coppola et al. [72] used CNTs as reinforcements in cement matrix to enhance the impact strength. Cement is used in heavy concrete structures which need to be designed to resist heavy impact loads as they are extensively used in military structures, nuclear reactor containment, etc. Interestingly, CNTs alone did not enhance the strength of mortar but when CNTs were added in combination with glass reinforced plastics (GRP), there was an increase in fracture energy and tensile strength. Split Hopkinson pressure bar (SHPB) was used for evaluation of dynamic strength at strain rates of about 150 s^{-1}.

Mechanical behavior was studied with electrical change in resistance by Lim et al. [73] using SHPB. Damage was observed for CNT/E-glass/SC-15 composite specimen under dynamic compressive load. This behavior was correlated with increased electrical resistance, and it occurs only after impacts that cause a decrease in stiffness of specimen.

Along with stress, it is very important to see extent of deformation the sample undergoes when it experiences heavy impact loads. The maximum strain that a material undergoes gives a measure of the energy required for the material to undergo permanent fracture; in other words, it gives the measure of its toughness. For any practical shock load-related application, the specimen which undergoes such heavy impact must be capable of absorbing the load and its deformation also becomes important to analyze its repeated use. Shielding specimen can be one-time use substances, or they can be used multiple times also depending upon their state after first-time use. However, dynamic impact loads generally cause a permanent strain or deformation in the specimen. Moreover, various composites exhibiting properties to absorb heavy impact loads must also be studied for the maximum strain they undergo before failing because if the space where that specimen is installed does not permit this deformation or extension, then it can escalate the damage by interfering with other parts of the host machinery. So it becomes imperative to evaluate the dimensional change caused in samples also.

For low-strain-rate range, Weeks et al. [74] have shown that as strain rate increases from 10 to 100,000 μs^{-1} on AS4 thermoplastics off axis composites, the stress–strain behavior shows a nonlinear behavior. The maximum strain varies from 1.7 to 2.3 % in the above strain range. Shan et al. [75] used the specimen as granite and marble and performed impact tests using SHPB. It was found that stress–strain behavior was linear for a small strain range, and beyond that, a nonlinear pattern was reported. The high strain rate was measured at high impact velocity, and as velocity increased, the maximum strain also increased from 100×10^{-4} % to 250×10^{-4} %. Various elastic materials like silicon rubber and AA6005-T6 were tested using SHPB by Ramezani and Ripin [76]. They reported the stress–strain pattern for low strain rates of 50–400 s^{-1}. The maximum stress for this specimen kept increasing with increase in strain rate, and maximum strain also increased from 0.6 to 0.8 % for silicon rubber. Tension test using SHPB was performed on polycarbonate for strain rates in the range of 0.001–1,700 s^{-1} by Cao et al. [10], and again an increase of maximum strain and dynamic strength with increase in strain rate was reported. Siviour et al. [77] performed compression impact tests on polycarbonates and polyvinylidene diflouride for strain rates of between 10^{-4} and $10^4 \, s^{-1}$ at different temperatures. Maximum strains and maximum stresses for these samples also increased with strain rate.

Hence, it has been observed that SHPB has been the most widely used instrument for high-strain-rate load testing of materials.

References

1. M.S.P. Shaffer, J.K.W. Sandler, Carbon nanotube/nanofibre polymer composites. *Processing and properties of nanocomposites* (World Scientific, Singapore, 2006), pp. 1–59
2. S. Iijima, Helical microtubules of graphitic carbon. Nature **354**(6348), 56–58 (1991)
3. R.S. Ruoff, D. Qian, W.K. Liu, Mechanical properties of carbon nanotubes: theoretical predictions and experimental measurements. Comptes. Rendus Phys. **4**(9), 993–1008 (2003)
4. M.S. Dresselhaus, G. Dresselhaus, P.C. Eklund, *Science of Fullerenes and Carbon Nanotubes: Their Properties and Applications* (Academic Press, 1996)
5. A. Sears, R. Batra, Macroscopic properties of carbon nanotubes from molecular-mechanics simulations. Phys. Rev. B **69**(23), 235406 (2004)
6. T.W. Ebbesen, *Carbon Nanotubes: Preparation and Properties* (Taylor and Francis, 1996)
7. W. Yu, W.X. Xi, N. Xianggui, Atomistic simulation of the torsion deformation of carbon nanotubes. Model. Simul. Mater. Sci. Eng. **12**(6), 1099 (2004)
8. G. Gao, T. Cagin, W.A. Goddard III, Energetics, structure, mechanical and vibrational properties of single-walled carbon nanotubes. Nanotechnology **9**(3), 184 (1998)
9. M.S. Dresselhaus, G. Dresselhaus, J.C. Charlier, E. Hernández, Electronic, thermal and mechanical properties of carbon nanotubes. Philos. Trans. A. Math. Phys. Eng. Sci. **362**(1823), 2065–2098 (2004)
10. K. Cao, X. Ma, B. Zhang, Y. Wang, Y. Wang, Tensile behavior of polycarbonate over a wide range of strain rates. Mater. Sci. Eng. A **527**(16–17), 4056–4061 (2010)
11. S. Berber, Y.-K. Kwon, D. Tománek, Unusually high thermal conductivity of carbon nanotubes. Phys. Rev. Lett. **84**(20), 4613–4616 (2000)
12. E. Pop, D. Mann, Q. Wang, K. Goodson, H. Dai, Thermal conductance of an individual single-wall carbon nanotube above room temperature. Nano Lett. **6**(1), 96–100 (2006)

13. C. Goze, L. Vaccarini, L. Henrard, P. Bernier, E. Hemandez, A. Rubio, Elastic and mechanical properties of carbon nanotubes. Synth. Met. **103**(1), 2500–2501 (1999)
14. G. Dereli, C. Özdoğan, Structural stability and energetics of single-walled carbon nanotubes under uniaxial strain. Phys. Rev. B **67**(3), 35416 (2003)
15. S. Gupta, K. Dharamvir, V.K. Jindal, Elastic moduli of single-walled carbon nanotubes and their ropes. Phys. Rev. B **72**(16), 165428 (2005)
16. K. Mylvaganam, L.C. Zhang, Important issues in a molecular dynamics simulation for characterising the mechanical properties of carbon nanotubes. Carbon **42**(10), 2025–2032 (2004)
17. N.R. Raravikar, P. Keblinski, A.M. Rao, M.S. Dresselhaus, L.S. Schadler, P.M. Ajayan, Temperature dependence of radial breathing mode Raman frequency of single-walled carbon nanotubes. Phys. Rev. B **66**(23), 235424 (2002)
18. V.K. Jindal, J. Kalus, Calculation of thermal expansion and phonon frequency shift in deuterated naphthalene. Phys. Status Solidi **133**(1), 89–99 (1986)
19. P. Jindal, V.K. Jindal, Model for compression of fullerenes and carbon nanotubes. Mol. Simul. **31**(12), 807–810 (2005)
20. P. Jindal, V.K. Jindal, Strains in axial and lateral directions in carbon nanotubes. J. Comput. Theor. Nanosci. **3**(1), 148–152 (2006)
21. B. Yakobson, R. Smalley, Fullerene Nanotubes: C 1,000,000 and Beyond Some unusual new molecules—long, hollow fibers with tantalizing electronic and mechanical properties—have joined diamonds and graphite in the carbon family. Am. Sci. **85**, 324–337 (1997)
22. B.I. Yakobson, Mechanical relaxation and 'intramolecular plasticity' in carbon nanotubes. Appl. Phys. Lett. **72**(8), 918 (1998)
23. H. Postma, M. de Jonge, Z. Yao, C. Dekker, Electrical transport through carbon nanotube junctions created by mechanical manipulation. Phys. Rev. B **62**(16), R10653–R10656 (2000)
24. M. Sammalkorpi, A. Krasheninnikov, A. Kuronen, K. Nordlund, K. Kaski, Mechanical properties of carbon nanotubes with vacancies and related defects. Phys. Rev. B **70**(24), 245416 (2004)
25. H.M. Cheng, F. Li, G. Su, H.Y. Pan, L.L. He, X. Sun, M.S. Dresselhaus, Large-scale and low-cost synthesis of single-walled carbon nanotubes by the catalytic pyrolysis of hydrocarbons. Appl. Phys. Lett. **72**(25) (1998)
26. M. Endo, K. Takeuchi, S. Igarashi, K. Kobori, M. Shiraishi, H.W. Kroto, The production and structure of pyrolytic carbon nanotubes (PCNTs). J. Phys. Chem. Solids **54**(12), 1841–1848 (1993)
27. R. Andrews, D. Jacques, D. Qian, T. Rantell, Multiwall carbon nanotubes: synthesis and application. Acc. Chem. Res. **35**(12), 1008–1017 (2002)
28. T. Guo, P. Nikolaev, A. Thess, D.T. Colbert, R.E. Smalley, Catalytic growth of single-walled nanotubes by laser vaporization, vol. 2614 Sept, 1995
29. J. Zhu, J. Kim, H. Peng, J.L. Margrave, V.N. Khabashesku, E.V. Barrera, Improving the dispersion and integration of single-walled carbon nanotubes in epoxy composites through functionalization. Nano Lett. **3**(8), 1107–1113 (2003)
30. M.S.P. Shaffer, X. Fan, A.H. Windle, Dispersion and packing of carbon nanotubes. Carbon **36**(11), 1603–1612 (1998)
31. P. Pötschke, T.D. Fornes, D.R. Paul, Rheological behavior of multiwalled carbon nanotube/polycarbonate composites. Polymer (Guildf) **43**(11), 3247–3255 (2002)
32. M. Moniruzzaman, K.I. Winey, R.V. April, V. Re, V. Recei, V. May, Polymer nanocomposites containing carbon nanotubes. Macromolecules **39**(16), 5194–5205 (2006)
33. M.S.P. Shaffer, A.H. Windle, Fabrication and characterization of carbon nanotube/poly(vinyl alcohol) composites. Adv. Mater. **11**(11), 937–941 (1999)
34. M. Sennett, E. Welsh, J.B. Wright, W.Z. Li, J.G. Wen, Z.F. Ren, Dispersion and alignment of carbon nanotubes in polycarbonate. Appl. Phys. A **76**(1), 111–113 (2003)
35. M. Cochet, W.K. Maser, A.M. Benito, M.A. Callejas, M.T. Martínez, J.-M. Benoit, J. Schreiber, O. Chauvet, Synthesis of a new polyaniline/nanotube composite: 'in-situ' polymerisation and charge transfer through site-selective interaction. Chem. Commun. **16**, 1450–1451 (2001)

36. S. Stankovich, D.A. Dikin, G.H.B. Dommett, K.M. Kohlhaas, E.J. Zimney, E.A. Stach, R.D. Piner, S.T. Nguyen, R.S. Ruoff, Graphene-based composite materials. Nature **442**(7100), 282–286 (2006)

37. J.K.W. Sandler, S. Pegel, M. Cadek, F. Gojny, M. Van Es, J. Lohmar, W.J. Blau, K. Schulte, A.H. Windle, M.S.P. Shaffer, A comparative study of melt spun polyamide-12 fibres reinforced with carbon nanotubes and nanofibres. Polymer (Guildf) **45**(6), 2001–2015 (2004)

38. S. Kanagaraj, F.R. Varanda, T.V. Zhil'tsova, M.S.A. Oliveira, J.A.O. Simões, Mechanical properties of high density polyethylene/carbon nanotube composites. Compos. Sci. Technol. **67**(15–16), 3071–3077 (2007)

39. K. Prashantha, Multi-walled carbon nanotube filled polypropylene nanocomposites based on masterbatch route: improvement of dispersion and mechanical properties through PP-g-MA addition. Express. Polym. Lett. **2**(10), 735–745 (2008)

40. M.C. Weisenberger, E.A. Grulke, D. Jacques, A.T. Rantell, R. Andrewsa, Enhanced mechanical properties of polyacrylonitrile/multiwall carbon nanotube composite fibers. J. Nanosci. Nanotechnol. **3**(6), 535–539 (2003)

41. M. Cadek, J.N. Coleman, V. Barron, K. Hedicke, W.J. Blau, Morphological and mechanical properties of carbon-nanotube-reinforced semicrystalline and amorphous polymer composites. Appl. Phys. Lett. **81**(27), 5123–5125 (2002)

42. D. Qian, E.C. Dickey, R. Andrews, T. Rantell, Load transfer and deformation mechanisms in carbon nanotube-polystyrene composites. Appl. Phys. Lett. **76**(20), 2868 (2000)

43. R. Andrews, D. Jacques, M. Minot, T. Rantell, Fabrication of carbon multiwall nanotube/polymer composites by shear mixing. Macromol. Mater. Eng. **287**(6), 395–403 (2002)

44. A.A. Mamedov, N.A. Kotov, M. Prato, D.M. Guldi, J.P. Wicksted, A. Hirsch, Molecular design of strong single-wall carbon nanotube/polyelectrolyte multilayer composites. Nat. Mater. **1**(3), 190–194 (2002)

45. H. Cai, F. Yan, Q. Xue, Investigation of tribological properties of polyimide/carbon nanotube nanocomposites. Mater. Sci. Eng. A **364**(1–2), 94–100 (2004)

46. E.J. Siochi, D.C. Working, C. Park, P.T. Lillehei, J.H. Rouse, C.C. Topping, A.R. Bhattacharyya, S. Kumar, Melt processing of SWCNT-polyimide nanocomposite fibers. Compos. Part B Eng. **35**(5), 439–446 (2004)

47. X. Li, W. Guan, H. Yan, L. Huang, Fabrication and atomic force microscopy/friction force microscopy (AFM/FFM) studies of polyacrylamide–carbon nanotubes (PAM–CNTs) copolymer thin films. Mater. Chem. Phys. **88**(1), 53–58 (2004)

48. Y.-H. Liao, O. Marietta-Tondin, Z. Liang, C. Zhang, B. Wang, Investigation of the dispersion process of SWNTs/SC-15 epoxy resin nanocomposites. Mater. Sci. Eng. A **385**(1), 175–181 (2004)

49. J.N. Coleman, U. Khan, W.J. Blau, Y.K. Gun'ko, Small but strong: a review of the mechanical properties of carbon nanotube–polymer composites. Carbon **44**(9), 1624–1652 (2006)

50. Y. Xu, Q. Li, D. Sun, W. Zhang, G.-X. Chen, A strategy to functionalize the carbon nanotubes and the nanocomposites based on poly(L-lactide). Ind. Eng. Chem. Res. **51**(42), 13648–13654 (2012)

51. P.R. Mantena, A.H.D. Cheng, A. Al-ostaz, A.M. Rajendran, Blast and impact resistant composite structures for navy ships, in *Proceedings of the ONR Solid Mechanics Program—Marine Composites and Sandwich Structures* ed. by Ahmed Al-Ostaz (2009) pp. 417–426

52. F.J. Yang, W.J. Cantwell, Impact damage initiation in composite materials. Compos. Sci. Technol. **70**(2), 336–342 (2010)

53. B. Das, K. Eswar Prasad, U. Ramamurty, C.N.R. Rao, Nano-indentation studies on polymer matrix composites reinforced by few-layer graphene. Nanotechnology **20**(12), 125705 (2009)

54. S.R.C. Vivekchand, U. Ramamurty, C.N.R. Rao, Mechanical properties of inorganic nanowire reinforced polymer–matrix composites. Nanotechnology **17**(11), S344–S350 (2006)

55. S.J. Bull, Analysis methods and size effects in the indentation fracture toughness assessment of very thin oxide coatings on glass. Comptes. Rendus Mécanique **339**(7–8), 518–531 (2011)

56. M. Ehsani, H.A. Khonakdar, A. Ghadami, Assessment of morphological, thermal, and viscoelastic properties of epoxy vinyl ester coating composites: role of glass flake and mixing method. Prog. Org. Coatings **76**(1), 238–243 (2013)

57. S.L. Gao, E. Mäder, R. Plonka, Nanostructured coatings of glass fibers: Improvement of alkali resistance and mechanical properties. Acta Mater. **55**(3), 1043–1052 (2007)
58. M. Grujicic, B. Pandurangan, T. He, B.A. Cheeseman, C.-F. Yen, C.L. Randow, Computational investigation of impact energy absorption capability of polyurea coatings via deformation-induced glass transition. Mater. Sci. Eng. A **527**(29–30), 7741–7751 (2010)
59. A. Nalbant, Ö. Ertek, İ. Okur, Producing CuO and ZnO composite thin films using the spin coating method on microscope glasses. Mater. Sci. Eng. B **178**(6), 368–374 (2013)
60. Y.P. Wu, M.O. Alam, Y.C. Chan, B.Y. Wu, Dynamic strength of anisotropic conductive joints in flip chip on glass and flip chip on flex packages. Microelectron. Reliab. **44**(2), 295–302 (2004)
61. J.O. Aguilar, O. Gomez-Daza, L.A. Brito, M.T.S. Nair, P.K. Nair, Optical and mechanical characteristics of clear and solar control laminated glass using zinc sulphide and copper sulphide thin films. Surf. Coatings Technol. **200**(7), 2557–2565 (2005)
62. C.A. Mitchell, J.L. Bahr, S. Arepalli, J.M. Tour, R. Krishnamoorti, Dispersion of functionalized carbon nanotubes in polystyrene. Macromolecules **35**(23), 8825–8830 (2002)
63. S. Kumar, R. Kumar, V.K. Jindal, L.M. Bharadwaj, Immobilization of single walled carbon nanotubes on glass surface. Mater. Lett. **62**(4), 731–734 (2008)
64. S. Flink, F.C.J.M. van Veggel, D.N. Reinhoudt, Functionalization of self-assembled monolayers on glass and oxidized silicon wafers by surface reactions. J. Phys. Org. Chem. **14**(7), 407–415 (2001)
65. M.A. Kaiser, A. Wicks, L. Wilson, W. Saunders, Advancements in the Split Hopkinson bar test (1998)
66. R. Khare, S. Bose, Carbon nanotube based composites—a review. J. Miner. Mater. Charact. Eng. **4**(1), 31–46 (2005)
67. P.J.F. Harris, Carbon nanotube composites. Int. Mater. Rev. **49**(1), 31–43 (2004)
68. M. Hosur, J. Alexander, U. Vaidya, S. Jeelani, High strain rate compression response of carbon/epoxy laminate composites. Compos. Struct. **52**(3–4), 405–417 (2001)
69. A. Jadhav, E. Woldesenbet, S.-S. Pang, High strain rate properties of balanced angle-ply graphite/epoxy composites. Compos. Part B Eng. **34**(4), 339–346 (2003)
70. W. Chen, F. Lu, M. Cheng, Tension and compression tests of two polymers under quasi-static and dynamic loading. Polym. Test. **21**(2), 113–121 (2002)
71. M. Zappalorto, M. Salviato, M. Quaresimin, Influence of the interphase zone on the nanoparticle debonding stress. Compos. Sci. Technol. **72**(1), 49–55 (2011)
72. L. Coppola, E. Cadoni, D. Forni, A. Buoso, Mechanical characterization of cement composites reinforced with fiberglass, carbon nanotubes or glass reinforced plastic (GRP) at high strain rates. Appl. Mech. Mater. **82**, 190–195 (2011)
73. A.S. Lim, Q. An, T.-W. Chou, E.T. Thostenson, Mechanical and electrical response of carbon nanotube-based fabric composites to Hopkinson bar loading. Compos. Sci. Technol. **71**(5), 616–621 (2011)
74. C.A. Weeks, C.T. Sun, C.A. Weeksa, C.T. Sud*, Compos. Sci. Technol. 58, **3538**(97), 603–611 (1998)
75. R. Shan, Y. Jiang, B. Li, Obtaining dynamic complete stress–strain curves for rock using the Split Hopkinson pressure bar technique. Int. J. Rock Mech. Min. Sci. **37**(6), 983–992 (2000)
76. M. Ramezani, Z.M. Ripin, Combined experimental and numerical analysis of bulge test at high strain rates using split Hopkinson pressure bar apparatus. J. Mater. Process. Technol. **210**(8), 1061–1069 (2010)
77. C.R. Siviour, S.M. Walley, W.G. Proud, J.E. Field, The high strain rate compressive behaviour of polycarbonate and polyvinylidene difluoride. Polymer (Guildf) **46**(26), 12546–12555 (2005)

Chapter 2
Fabrication of Composites and Coatings

Among the thermoplastic group of polymers, polycarbonates (PC) have attracted a great deal of attention due to their ability to be easily worked upon and moldability. Figure 2.1 gives the general structure of PC. Their capability to resist temperature and impact makes them a common application material in housewares, laboratories, and industries. Therefore, any modification in their properties to suite specific requirements becomes interesting.

Polycarbonates are lightweight and smooth materials. Their softness and low scratch resistance makes them vulnerable for certain load-related applications. Some of their mechanical properties have been given in Table 2.1. Polycarbonate has a glass transition temperature of about 150 °C. When heated beyond this point, it softens gradually and begins to flow above 300 °C. PC undergoes large plastic deformations and does not break easily, which is a suitable property for any engineering application.

Other polymers like polymethyl acrylate (PMMA) are brittle and suddenly break even at room temperature. Lot of enclosures, bottles, glass cases, lenses, etc., are fabricated using PC. Polypropylene (PP) does not shrink uniformly so it is difficult to achieve accurate fabrication.

Just like polymers, glass has been widely used over the years for many applications. Window panes are one of the commonest applications of glass. Major component of glass is silica. Glass is of various types like soda lime, borosilicate, optical glass, etc. Their Young's modulus varies between 40 and 75 GPa, and Vickers hardness number is about 700 Vickers. Despite its various uses, it has some limitations. Glass is very brittle and susceptible to heavy loading. An attempt to modify its brittle behavior and increase its load-bearing capacity needs to be explored either by embedding strong fibers inside or by coating it. Borosilicate is a type of glass that is widely used for laboratory glassware. Some properties of borosilicate glass are given in Table 2.2. Borosilicate glass has excellent thermal properties with its low coefficient of expansion and high softening point.

© The Author(s) 2015
P. Jindal, *High Strain Rate Behavior of Nanocomposites and Nanocoatings*,
SpringerBriefs in Materials, DOI 10.1007/978-3-319-14481-8_2

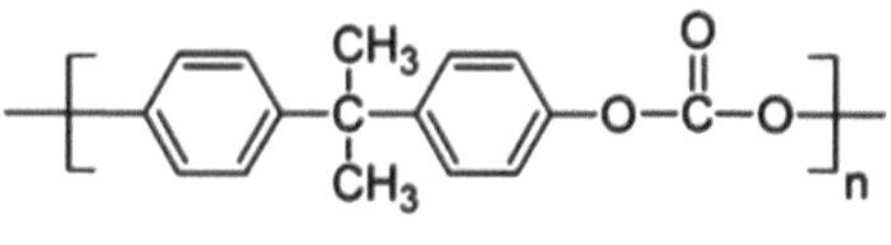

Fig. 2.1 General structure of a PC molecule

Table 2.1 Properties of polycarbonates

Property	Value
Young's modulus	2.0–2.4 GPa
Tensile strength	55–75 MPa
Poisson's ratio	0.37
Thermal conductivity at 23 °C	0.19–0.22 W/mK
Density	1,200 kg/m^3

Table 2.2 Properties of borosilicate glass

Property	Value
Density	22.3 kg/m^3
Specific heat (20 °C)	750 J/kg °C
Thermal conductivity (20 °C)	1.14 W/m °C
Poisson's ratio (25–400 °C)	0.2
Young's modulus (25 °C)	64 GPa

Resistance to acid, solvents, and other solutions is also high although some acids can cause rapid corrosion also. In several defense vehicles, glass window panes are used. Toughened glass is one of the strong materials, but it is very heavy in weight. So, for protecting man power inside the vehicle, the glass itself can be made strong enough to resist the impact or it can be coated with some material which can resist the impact and in turn save the glass from breakage. Coatings on glass can cause loss in transparency of the glass to some extent, but depending upon the overall application and extent of transparency required, the choice of coating materials can be explored.

Therefore, both borosilicate glass and PC have been used as base materials to explore their mechanical properties with MWCNTs as coat and fillers.

2.1 Composite Fabrication with PC and MWCNT

Composite materials of PC and MWCNTs were fabricated to study the role of MWCNTs in enhancing the mechanical properties of PC-based composites. As-synthesized MWCNTs (a-MWCNT) with diameter about 10–30 nm and length 1–10 μm were used as filler material, and PC was used as base matrix. Figure 2.2 shows Raman spectra of the pure MWCNTs which indicates signature peaks as D (1,338/cm), G (1,563/cm), and G′ (2,678/s).

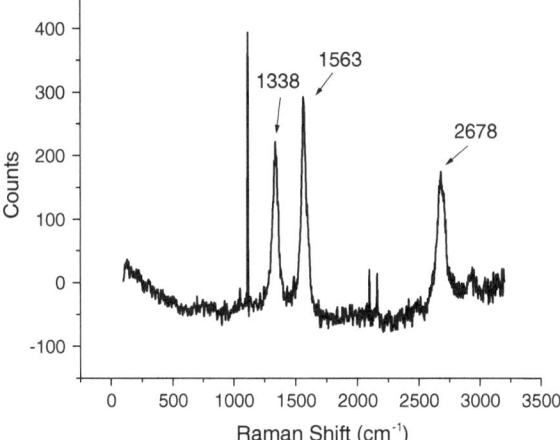

Fig. 2.2 Raman spectra for MWCNTs with D (1,338/cm) peak, G (1,563/cm), and G′ (2,678/s) peaks

After characterization of the MWCNTs as mentioned above for the fabrication of MWCNT-PC composites [1], MWCNTs were first ultrasonically dispersed in chloroform to obtain a stable suspension of carbon nanotubes. Clear solutions of PC were also formed in chloroform by stirring. Chloroform was used the solvent in a quantity about 10 times MWCNTs and PC. The suspensions were then mixed with solutions of PC in chloroform to obtain a series of mixtures of MWCNT/PC containing different weight percent of MWCNT with compositions 0.5, 0.75, 2.0, 5.0, and 10.0 wt% in PC to relate the effect of amount of MWCNTs on mechanical strength. The mixtures were then stirred to obtain uniform dispersions of MWCNTs in PC. Thin polymer films were cast from this solution by pouring the solution into glass petri dish and allowing the solvent to evaporate. The resulting films had a thickness of about 0.2 mm. These films were then broken into smaller pieces and inserted in the die of a compression molding machine.

They were heated to temperatures of about 100 °C and then manually pressed in the compression molding machine to get disk-shaped samples of 10 mm diameter and 5 mm thickness. As per the specifications required for high-strain-rate loading, minor finishing processes were done on the sample surface to get smooth surface finish for load testing purposes. For characterization of the composites, Raman spectroscopy was done as shown in Fig. 2.3 which indicates the comparisons of D and G peaks (as discussed for Fig. 2.2) for composites consisting of 5 and 10 wt% MWCNT compositions.

It was observed that the intensity for D and G peaks is different for both composites. For 5 % MWCNT composition, the intensity for D (1,343/cm) peak is 12,139.6, while for G (1,562/cm) peak, it is 12,722.6. Similarly, for 10 % MWCNT composition, the intensity for D (1,361/cm) peak is 20,829.7, while for G (1,573/cm) peak, it is 22,392.6. So, it can be observed that the intensities of 10 % MWCNTs composite for both D and G peaks are almost twice that of 5 % MWCNTs composite which is in coherence with the composition of MWCNTs in the composite.

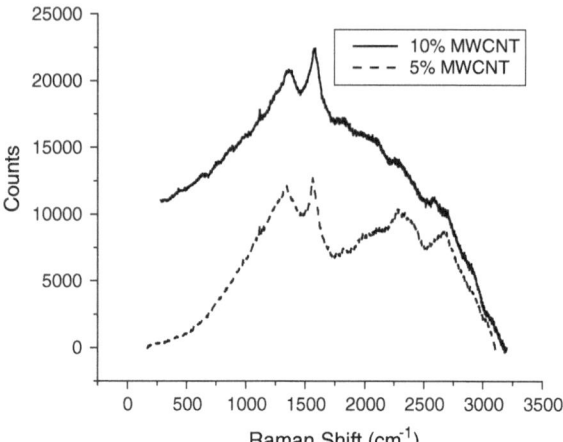

Fig. 2.3 Raman spectra for composite with 5 % MWCNT and 10 % MWCNT composition in PC

Therefore, 10 specimens were prepared for every composition to initiate the process of mechanical characterization of these composites.

2.2 Coating Glass with MWCNTs

Borosilicate glass pieces of disk shape having diameter 10 mm and thickness 5 mm were procured and used as the base material to be coated with MWCNTs. Their surfaces were cleaned with ethanol. MWCNTs of variable amounts were mixed with dimethylformamide (DMF) and ultrasonicated for a few hours to ensure reasonable dispersion. Measured quantities of different compositions of these MWCNT solutions were then spread over the surfaces glass pieces [2] to form uniform MWCNT coating on the glass surface. The different concentrations of these MWCNT solutions and amount spread over the glass pieces are given in Table 2.3. On evaporation of the solvent, coatings of varied thickness and quantity of MWCNTs distributed reasonably uniformly as solvent on the surface of glass samples of 79-mm^2 cross-sectional surface area corresponding to 10 mm diameter was obtained.

A simple estimate of a single layer of average thickness T of MWCNTs of bulk density ρ when spread over a surface area A of the glass disk will have mass as $m = \rho AT$. So, the thickness can be found by the equation

$$T = \frac{m}{\rho A}$$

The average bulk density of the procured MWNCTs was 100 mg/cm^3, average length = 5 μm. Glass surface area A = 0.79 cm^2 and mass of MWCNTs m = 0.1–0.8 mg deposited on the surface meant that the thickness for the samples varied from 10 to 100 μm.

Table 2.3 Samples of various concentrations of MWCNT solution on glass, volume of poured solution, thickness of coat, and rough estimate of number of layers

Sample no.	Concentration of coating on glass (mg/μL)	Quantity of solution poured (μL)	Coating thickness (μm)	Estimated no. of layers
1	10/1,000	10	12	3
2	18.6/930	10	25	5
3	20/520	10	49	10
4	25/400	10	80	16
5	15/200	10	95	19

It also meant that the samples were coated with 5–20 layers of MWCNTs. This way the MWCNT layers can be controlled to about 50 by varying the deposits of MWCNTs even if the MWCNTs are vertically aligned. The data of estimated number of layers are also presented in Table 2.3. It may be noted that the number of layers is based upon the assumption that MWCNTs are vertically aligned; however, in reality, MWCNTs can be a combination of various alignments. Hence, the number of layers given is a lower estimate.

In this way, 10 specimens of each coated composition specimen were prepared.

Hence, following simple fabrication procedures, PC composites with different MWCNT compositions and borosilicate glass specimen with variable coating thickness of MWCNTs were fabricated as per specifications of high-strain rate testing experimental setup.

References

1. S. Pande, R. Mathur, Synthesis and characterization of multiwalled carbon nanotubes-polymethyl methacrylate composites prepared by in situ polymerization method. Polym. Compos. **30**(9), 1312–1317 (2009)
2. A. Ma, J. Lu, S. Yang, K.M. Ng, Quantitative non-covalent functionalization of carbon nanotubes. J. Clust. Sci. **17**(4), 599–608 (2006)

Chapter 3
High-Strain-Rate Loading

The disk-shaped composite specimens with different compositions of MWCNTs were then subjected to high-strain-rate impact, and their dynamic impact loading behavior results were compared with variations in compositions of MWCNTs. The variation parameter here was only the composition of MWCNTs and not the geometry or orientation of the inner structure of specimen.

This setup for SHPB was used for compression testing, and it comprised of two high-strength elastic pressure bars in which the specimen was sandwiched. The pressure bars were made of high-strength maraging steel with yield strength ~1,750 MPa, diameter 20 mm, and length 2,000 mm. The projectile diameter was 20 mm, and length was 300 mm. Strain gauges of 120 Ω, 90° tee rosette precision strain gauges designated as EA-06-125TM-120 were used. For wave shaping, a 1.5-mm OFHC copper wave shaper was used.

During the experiments, the equilibrium was achieved for small durations of impact. However, as indicated in an earlier study on acrylic, for polymeric materials, dynamic equilibrium may not be achieved, but results of above equations are applicable [1].

Thus, for every sample tested under the dynamic load, the graphs between stress and strain, strain rate with time, true strain with time and the wave pattern of incidence, transmission, and reflection along the cycle with time were obtained. The general representation of these graphs is given in Figs. 3.1, 3.2, and 3.3.

Figure 3.1 is the general wave pattern of incident, reflected, and transmitted wave which is generated during the complete cycle of striker hitting the incident bar, incident bar compressing the sample, and hitting the output bar. The pattern depicts that the transmitted wave is not generated in the initial portion when incident wave is at its peak. As explained in the working of SHPB, when striker hits the incident bar only incident wave is generated. Once the incident bar compresses the sample and hits the output bar, only then reflection and transmission of waves takes place. Figure 3.2 gives the variation of strain rate in the short time span of

© The Author(s) 2015   29
P. Jindal, *High Strain Rate Behavior of Nanocomposites and Nanocoatings*,
SpringerBriefs in Materials, DOI 10.1007/978-3-319-14481-8_3

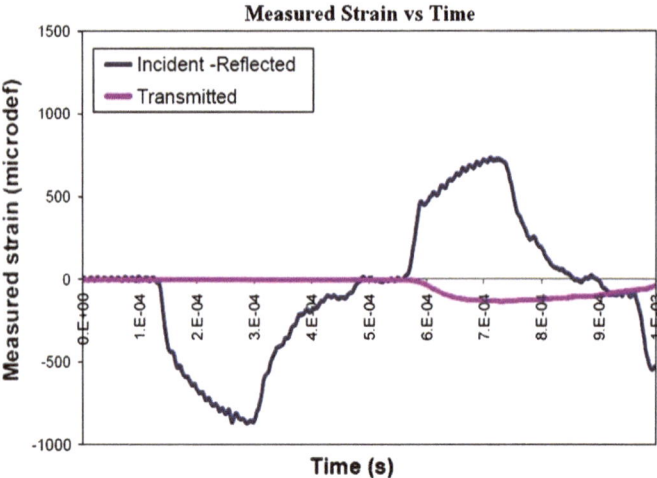

Fig. 3.1 Incident, reflected, and transmitted wave pattern for one complete cycle of dynamic load application for a sample

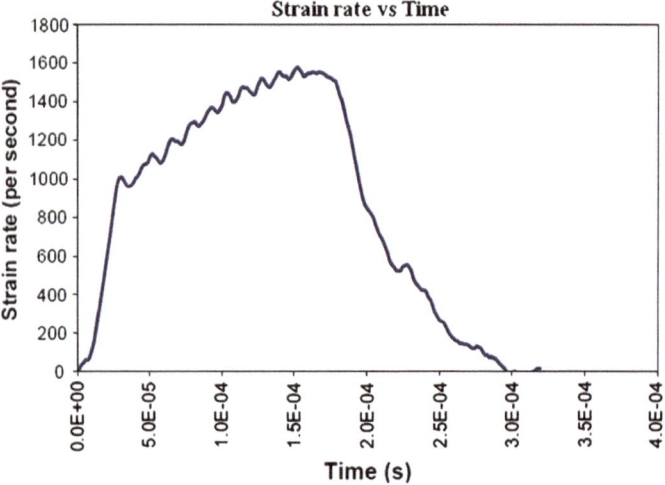

Fig. 3.2 Strain rate variation with time during one complete cycle of dynamic load application for a sample

the impact. Strain rate comes into picture immediately when the incident bar starts compressing the sample. The sample undergoes rapid compression, deformation, and ultimately fails. During this complete cycle, strain rate rises to a maximum level and then starts falling.

Figure 3.3 gives the strain or deformation behavior of the sample with time. As the ultimate condition of the sample is complete rupture, the strain shows a

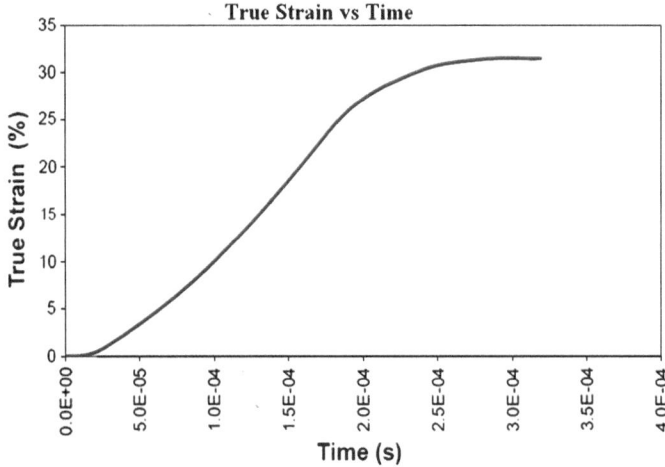

Fig. 3.3 Variation of true strain with time during one complete cycle of dynamic load application for a sample

consistent rise with time. This data are obtained for all the samples and are collectively used to generate the stress–strain behavior over the complete loading cycle for a particular specimen at a specific strain rate. Table 3.1 shows the summary of the maximum stress and strains obtained at various strain rates for every impact cycle on each sample composition.

For comparison purposes, the data for samples loaded between strain rates 2,000–2,800/s were analyzed because generally, the strain rate in the range of 2,000–2,800/s is a useful range in normal shock conditions encountered during aviation and defense-related applications [2]. So, compressive stress–strain behavior for various MWCNT–PC compositions for strain rates between 2,000 and 2,800/s is shown in Fig. 3.4.

It can be observed from Fig. 3.4 that minor compositions of MWCNTs enhance the maximum stress of the PC-based composites by nearly 10–20 % as compared to pure PC.

Maximum stress values of nearly 105 MPa were obtained for PC with MWCNT samples, while for pure PC samples, these were nearly 90 MPa. Although the data have been obtained for large values of strains, but for any practical applications, deformations/strains up to 20 % have been considered sufficient for failures. Hence, the effect of composition of MWCNTs on maximum stress for strains up to 20 % was analyzed. Therefore, Fig. 3.5 shows the effect of composition of MWCNTs on stress for specific strains of 5, 10, 15, and 20 %.

It can be observed that for low strains of 5 % the pure PC, 0.1 wt% MWCNT–PC, and 0.5 wt% MWCNT–PC composite resisted nearly the same stress. When the composition of MWCNTs was increased to 1 wt%, the maximum stress increased by about 20 %.

Table 3.1 Variation of elastic limit strain and maximum strain for different compositions at various strain rates

Strain (s⁻¹)	Elastic limit strain (%)	Maximum strain (%)
Pure PC sample		
2,361	5	50
2,677	7.5	60
0.1 % MWCNT–PC composite		
2,609	10	99
2,778	10	108
3,200	12	144
0.5 % MWCNT–PC composite		
2,032	8	66
2,186	10	73
2,768	10	108
2,845	10	113
1.0 % MWCNT–PC composite		
1,643	9	53
2,547	10	92
2,926	10	116
3,133	12	136

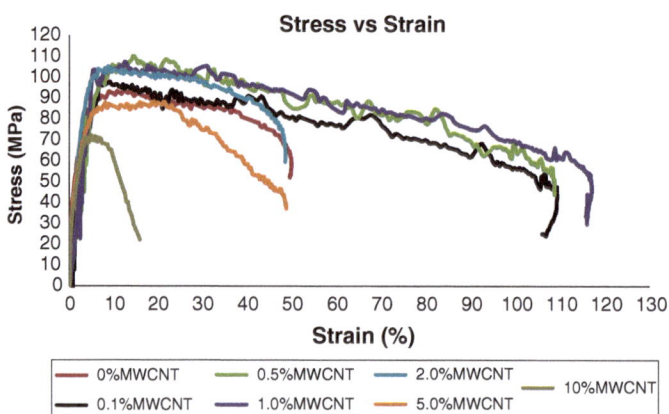

Fig. 3.4 True stress–strain curves for various compositions of MWCNT–PC composites at a strain rate of ~2,500/s

The maximum stress value corresponding to this increase was about 100 MPa for 1 wt% MWCNT–PC composite as compared to only 85 MPa for pure PC. An increase in composition to 2 wt% CNTs did not show any further increase in stress and remained almost constant. However, further increase in concentration of CNTs to 5 wt% resulted in a decrease in stress as it again fell to nearly the same value as that of pure PC.

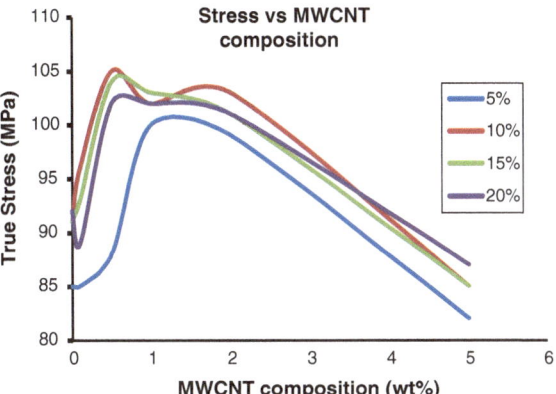

Fig. 3.5 Variation of stress resistance at a strain rate of ~2,500/s with varying concentration of MWCNTs in MWCNT–PC composites for different strains

Then, for higher strain of 10 %, there was again not much improvement in the stress for 0.1 % MWCNT–PC composite compared to pure PC (as in Fig. 3.5). For samples with 0.5 wt% CNT, the stress enhancement of about 15 % was observed. The stress increased from about 91 MPa in pure PC to about 105 MPa in the 0.5 wt% MWCNT–PC composite.

In the composition range of 0.5–2 wt% MWCNTs, the stress did not increase further. However, a further increase in CNT concentration to 5 wt% resulted in a decrease in stress values. Thus, maximum stress is observed in the range of 0.5–2 wt% MWCNT/PC composite.

As seen from the Fig. 3.5, this trend of increased stress with composition of MWCNTs continues for higher strains, i.e., 15 and 20 % also. This figure also indicates that the increased stress is offered by composites with 0.5–2 wt% MWCNTs only, and beyond this composition, the stress starts reducing. So, overall, for small deformations, the maximum resistance to dynamic load takes place for composites comprising 0.5–2 wt% of MWCNT in PC and beyond that, the resistance starts reducing. Two percentage can be considered as the critical limit of composition, beyond which further addition of MWCNTs does not enhance the impact strength. It is suggested that at higher concentration of MWCNTs aggregation of MWCNTs leads to lower impact strength of the samples. At low concentrations, the load-bearing capacity of the composite is enhanced because minor amounts of CNTs are well dispersed.

SEM images (Figs. 3.6, 3.7, 3.8, 3.9, and 3.10) were also obtained of the composite samples before and after the impact tests. The preimpact images in Figs. 3.6 and 3.7 show a regular topography. The images in Figs. 3.8, 3.9, and 3.10 indicate irregularity and a presence of MWCNTs bound with the base material. The well-dispersed CNTs in PC seem to hold the PC together depending upon concentration of CNTs.

Another aspect of this study relates the stress–strain behavior of a single composite sample at different strain rates.

Maximum and elastic limit strains which these samples underwent in the range of low strain rates (nearly 1,500/s) and high strain rates (nearly 3,000/s) have been

Fig. 3.6 SEM image
for 0.5 % MWCNT–PC
composite before impact
loading

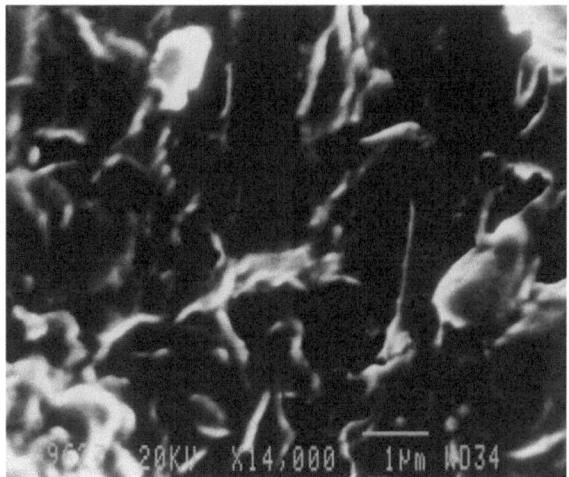

Fig. 3.7 SEM image
for 5.0 % MWCNT–PC
composite before impact
loading

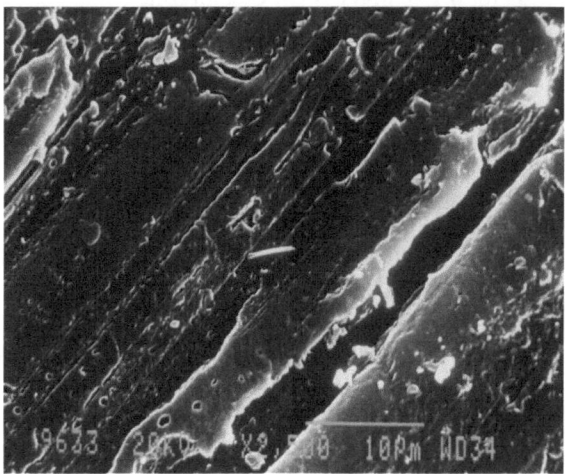

depicted in Table 3.1. It shows that both elastic limit strain and maximum strain increase for every sample concentration with increase in strain rate irrespective of the composition of the sample.

Figures 3.11, 3.12, and 3.13 indicate the stress–strain results for pure PC and small MWCNT compositions of 0.1 % MWCNT/PC and 0.5 % MWCNT/PC at different strain rates, respectively.

For higher compositions of 2 % MWCNT/PC and 5 % MWCNT/PC, stress–strain results under different strain rates are depicted graphically in Figs. 3.14 and 3.15. Whether the composition of MWCNTs is low or high, the pattern for maximum strain with strain rate remains similar. The maximum stress for each composition does not show any uniform variation with strain rate. However, if the focus is only

Fig. 3.8 **a**, **b** SEM images for 2 wt% MWCNT–PC composite after impact loading strain rate of 2,272/s at indicated magnifications

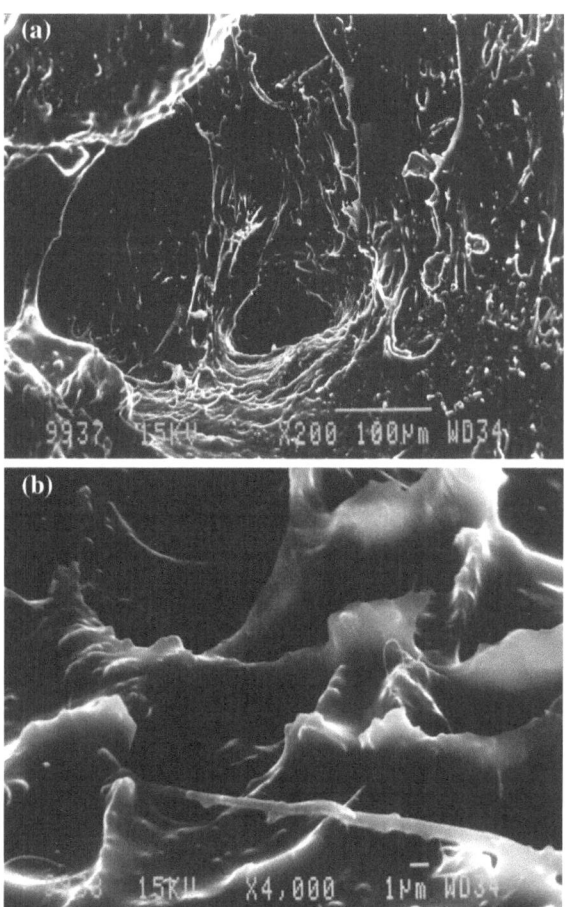

Fig. 3.9 SEM image for 5 wt% MWCNT–PC composite after impact loading strain rate of 2,040/s at indicated magnifications (×4,000, 1 µm)

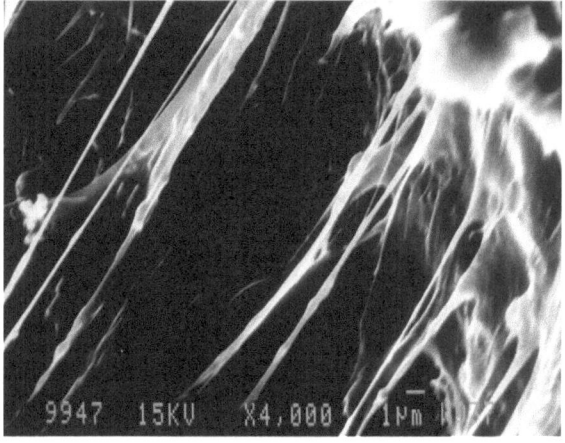

Fig. 3.10 SEM image
for 10 wt% MWCNT–PC
composite after impact
loading strain rate of 2,475/s
at indicated magnifications
(for ×1,000, 10 μm)

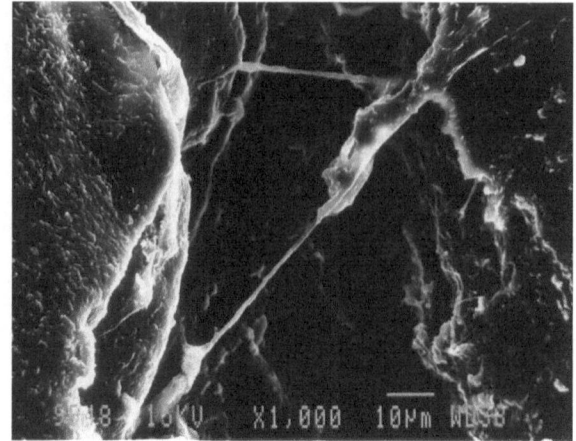

Fig. 3.11 Stress–strain
variation for pure PC under
different strain rates

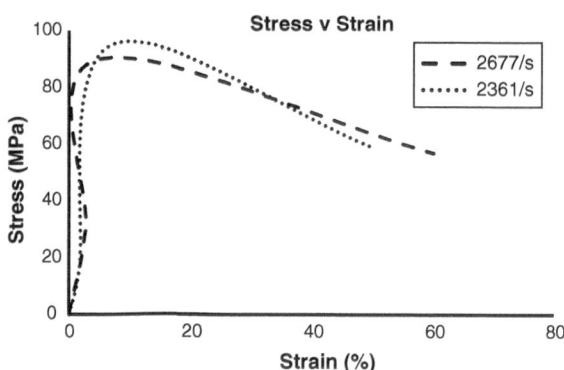

Fig. 3.12 Stress–strain
variation for 0.1 %
MWCNT–PC under different
strain rates

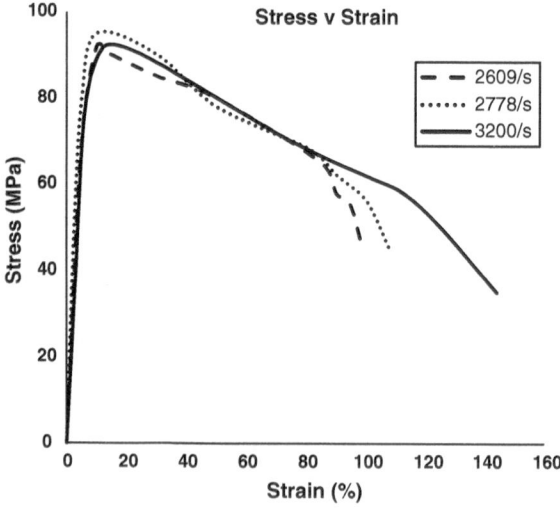

Fig. 3.13 Stress–strain variation for 0.5 % MWCNT–PC under different strain rates

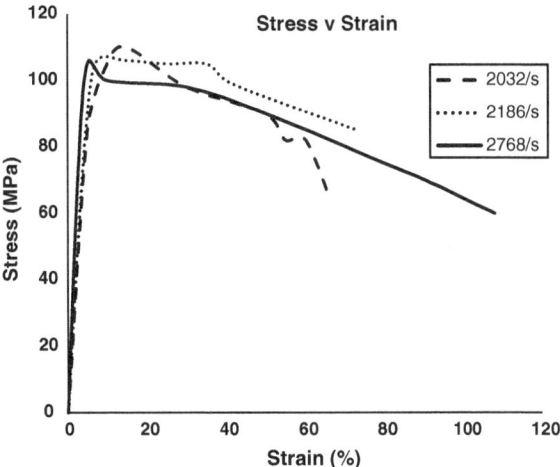

Fig. 3.14 Stress–strain variation for 2 % MWCNT–PC under different strain rates

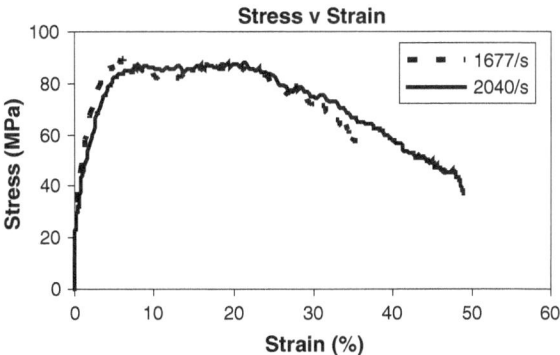

Fig. 3.15 Stress–strain variation for 5 % MWCNT–PC under different strain rates

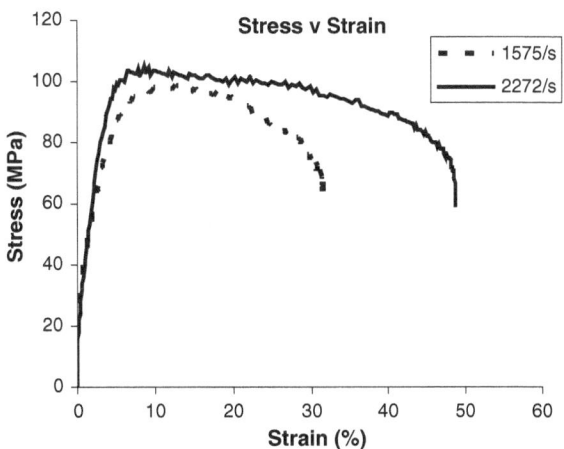

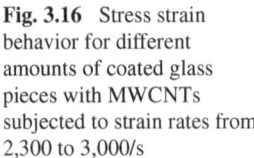

Fig. 3.16 Stress strain behavior for different amounts of coated glass pieces with MWCNTs subjected to strain rates from 2,300 to 3,000/s

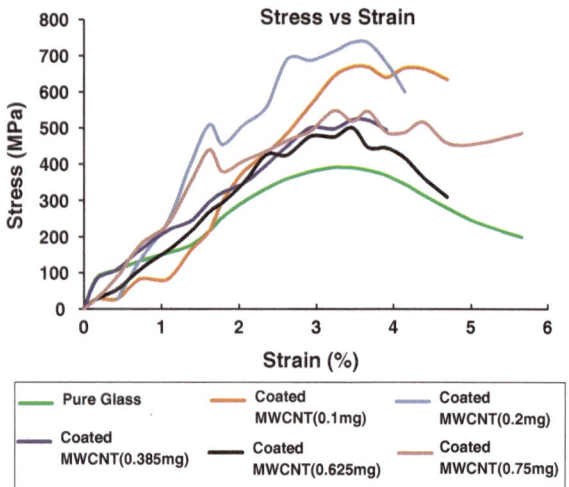

on deformation (strain), then it is observed that with increase in strain rate, the maximum strain for every composition increases before failure. This gives an estimate of the energy required for the specimen to deform permanently. As the strain rate increases, this energy also increases. The maximum strain provides important information in the context that for specific strain rate, the maximum deformation that the specimen undergoes can also be estimated. The final state of the specimen is in a broken or crushed condition after the impact loading process gets over. However, during deformation, its dimensional change may cause interference to various other devices installed around in its vicinity, which can escalate the damage caused by the impact.

In many situations, complex assemblies consisting of small elements are required to be protected from external dynamic loading. Materials which are able to resist stress (examined during SHPB dynamic load testing) can be used to protect such assemblies. However, even if these specimens are able to resist the stress, still they can cause damage to the assemblies due to the extent of final deformation state. The deformation of the specimens during impact can lead to interference with any of the small elements located in the neighborhood vicinity. Therefore, the study of maximum strain has to be coupled with the study of maximum stress and accordingly the positioning and location of the specimen can be decided owing to the maximum deformation which that specimen may undergo during a dynamic high-strain loading condition.

From Figs. 3.11, 3.12, 3.13, 3.14, and 3.15, it becomes evident that PC-based samples can deform to a large extent as the dynamic impact load increases, which is a similar behavior to various other materials like AS4 thermoplastics, granite, and marble [3–5].

High-strain-rate loading tests were also conducted on the MWCNT-coated glass specimens. Samples subjected to a strain rate of 2,300–3,000/s were compared for mechanical characterization. Compressive stress–strain behavior for glass pieces coated with MWCNTs of different deposits is shown in Fig. 3.16.

Fig. 3.17 Maximum stress
variation with different
coated MWCNTs–glass
samples subjected to strain
rates from 2,300 to 3,000/s

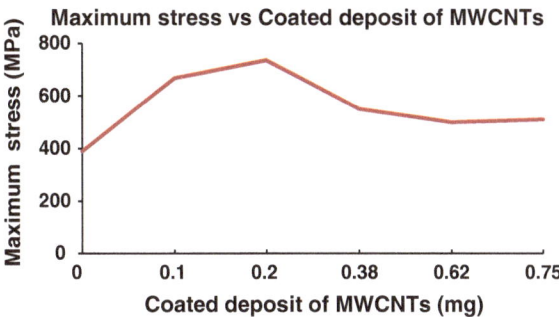

It is observed from Fig. 3.16 that a plastic deformation pattern is formed for all samples for a very small strain. Since glass is a brittle material, it does not undergo large dimensional deformation as in the case of polymer composites. Figure 3.16 shows that till a particular deposit of MWCNT coating, there is a substantial increase in the maximum stress, but after that it starts decreasing. Maximum stress resisted for pure glass is nearly 389 MPa. When this glass piece is non-covalently bonded with 0.1 mg of MWCNTs coating, this maximum stress limit reaches 667 MPa at nearly the same strain. Similarly, for 0.2-mg coating, the stress value is about 736 MPa. But beyond this, for coatings of 0.385, 0.625, and 0.75 mg, this maximum stress value remains nearly constant (500 MPa) which is still much higher than pure glass. So, in comparison with pure glass, the samples which were coated with very small amounts of 0.1 and 0.2 mg MWCNTs exhibited 50–70 % increase in maximum stress [6].

This coated deposit of MWCNTs has been related with thickness of coating earlier in Chap. 1. Hence, it also implies that a coating thickness of MWCNTs of about 12–25 μm is sufficient to enhance the stress absorption by almost 2 times. However, the improved degradation at higher concentration is most likely to be a result of slipping [7] of the layers among themselves as contact with glass gets lost. Coatings of nearly 0.4–0.8 mg mean that thickness of coatings reaches nearly 40–100 μm. So, the number of layers on the glass pieces increases accordingly.

The effect of variation in the deposit of MWCNT coatings on maximum stress as explained above is further depicted in Fig. 3.17.

3.1 Conclusions

High-strain-rate behavior under compressive loading was studied for different compositions of MWCNT–PC composites. The measurements of impact strength were done by using split Hopkinson pressure bar. The change in dynamic strength of polycarbonates due to varying concentration of carbon nanotubes was studied. It was observed that low concentrations of CNTs improve the impact strength of

PC samples by nearly 10–20 % as compared to pure PC. SEM images indicate that well-dispersed CNTs maintain their structure even after heavy impacts. Long lengths and proper dispersion of CNTs are suggested to be more important in resisting major part of the impact. Hence, depending upon the extent of expected deformation, the choice of composition can be made.

The role of strain rate on the stress–strain behavior of MWCNT–PC composites with specific reference to maximum strain of the specimen has also been studied. Irrespective of the fact that the pure PC is embedded with MWCNTs, the maximum strain always increases with the increase in strain rate. The deformation in the specimen when these specimens are subjected to such high impact loads also becomes a matter of concern. The final state of the specimen in an impact load is generally crushed or ruptured. During this deformation phase, the dimensions of specimen change and may cause interference to various other devices installed in its neighborhood vicinity, which can escalate the damage caused by the impact on that single specimen. So, even though a resilient specimen is installed for protection from heavy impact load, still it must conform to the space around it, and hence, its deformation must also be ascertained.

Similarly, glass samples which had MWCNT coatings of about 0.1 and 0.2 mg showed remarkable enhancement in the maximum stress absorbed as compared to pure glass. This increase was about 50–70 % in comparison with pure glass specimen. Maximum stress for 0.1 and 0.2 mg coating sample was nearly 689 and 736 MPa, respectively, while for pure glass maximum stress was 389 MPa. However, coatings of nearly 0.4, 0.6, and 0.8 mg did not show further increase. The maximum stress for these samples was nearly 500 MPa which was still about 30 % higher than pure glass but much less than 0.1 and 0.2 mg. The reason suggested for this reduction can be the increased thickness of coating that comprises of multiple layers of MWCNTs. As layers of coatings increase, there is slipping of these layers from the glass surface and among the layers themselves. As a result, the coatings slip away from the base glass surface and fail to offer higher resistance to impact.

It can be concluded that for the MWCNT–PC composites analyzed in the strain rate range of 2,000–2,800/s, MWCNT percentage of around 0.5–2 wt% represents an upper limit to enhance dynamic impact strength of PC composite. This strain rate is considered a useful range in normal shock conditions, encountered during aviation and defense requirements.

It also seems safe to conclude that small deposits of MWCNT coatings can protect the glass from high impact loads. To resist impact load under specific applications, an estimate of thickness of MWCNTs on glass surface can be done accordingly. So this can be useful as an impact stress sensor also. In fact, a stacking of multiple coated glass samples can be used to absorb desired impact as well as sensing unit for such impacts. As the glass piece was covered with minor amount of MWCNTs, the transparency loss was not significant.

References

1. N.K. Naik, Y. Perla, Mechanical behaviour of acrylic under high strain rate tensile loading. Polym. Test. **27**(4), 504–512 (2008)
2. M. Shazly, D. Nathenson, V. Prakash, Modeling of high-strain-rate deformation, fracture, and impact behavior of advanced gas turbine engine materials at low and elevated temperatures (Glenn Research Center, 2003)
3. F.J. Yang, W.J. Cantwell, Impact damage initiation in composite materials. Compos. Sci. Technol. **70**(2), 336–342 (2010)
4. C.A. Weeks, C.T. Sun, Modelling non-linear rate-dependent behaviour in fiber-reinforced composites. Compos. Sci. Technol. **58**, 603–611 (1998) (vol. 3538, no. 97)
5. R. Shan, Y. Jiang, B. Li, Obtaining dynamic complete stress–strain curves for rock using the split Hopkinson pressure bar technique. Int. J. Rock Mech. Min. Sci. **37**(6), 983–992 (2000)
6. P. Jindal, M. Goyal, N. Kumar, Dynamic impact absorption behaviour of glass coated with carbon nanotubes. J. Surf. Eng. Mater. Adv. Technol. **3**, 257–261 (2013)
7. M. Olek, K. Kempa, S. Jurga, M. Giersig, Nanomechanical properties of silica-coated multiwall carbon nanotubes-poly(methyl methacrylate) composites. Langmuir **21**(7), 3146–3152 (2005)

Chapter 4
Summary and Future

Summarizing this book, the methodology to fabricate composites of polycarbonates–multi-wall carbon nanotubes (PC-MWCNT) and fabricate protective coatings of multi-wall carbon nanotubes (MWCNT) layers on glass has been explained in detail. Realizing the significance of newly discovered materials of carbon nanotubes possessing highly attractive mechanical properties, there was a need to exploit these in modifying the easily available and workable low-cost engineering materials like polycarbonates (PC) and glass. This goal has been achieved by fabricating their composites and coatings and then characterizing these materials to ascertain whether the properties of these modified materials qualify as dynamic impact resistance substances. Such studies required the use of state-of-the-art experiments involving split Hopkinson pressure bar (SHPB) setups for dynamic loading.

For fabricating the composite, base matrix of PC was selected which is a polymer containing carbonate group and is of great commercial interest because PC is a soft and lightweight material; therefore, it can be easily worked upon, molded, and thermoformed. So the main goal was to modify PC by embedding modern materials like MWCNTs in it. Similarly, glass-based specimens were also modified by coating them with different compositions of MWCNT layers. These coatings could serve as excellent materials to resist impact loads and protect the glass surfaces. Simple chemical methods were used to fabricate composites and coated samples. Characterization techniques like scanning electron microscopy (SEM) and Raman spectrometry were also used to validate the composition of these composites. The properties of the composite and coated materials were extensively studied by SHPB.

Interestingly, these experiments indicated that minor compositions of nearly 0.5–2 wt% MWCNTs in PC were sufficient to enhance the dynamic strength by nearly 20 % in comparison with pure PC. Similarly, when a borosilicate glass surface of a unit cm^2 surface area was coated with only 0.1–0.2 mg MWCNTs, then the dynamic strength enhancement was nearly 50 % in comparison to a pure uncoated borosilicate glass surface (same area).

© The Author(s) 2015
P. Jindal, *High Strain Rate Behavior of Nanocomposites and Nanocoatings*,
SpringerBriefs in Materials, DOI 10.1007/978-3-319-14481-8_4

Depending upon the strength of the material and external impact load, various types of sensors and absorbers can be designed by varying the composition of MWCNTs as filler and coating specimen.

It is hoped that all the experimental work reported here will motivate the development of applications related to pressure sensing by using these composites which are of low cost and simple to produce. It is also hoped that more static load experimentation and theoretical work will be generated in the future to provide insight into the strength enhancement characteristics of the composites and coatings.